# International Series in Operations Research & Management Science



Volume 365

The book series **International Series in Operations Research and Management Science** encompasses the various areas of operations research and management science. Both theoretical and applied books are included. It describes current advances anywhere in the world that are at the cutting edge of the field. The series is aimed especially at researchers, advanced graduate students, and sophisticated practitioners.

The series features three types of books:

• Advanced expository books that extend and unify our understanding of particular areas.

• Research monographs that make substantial contributions to knowledge.

• Handbooks that define the new state of the art in particular areas. Each handbook will be edited by a leading authority in the area who will organize a team of experts on various aspects of the topic to write individual chapters. A handbook may emphasize expository surveys or completely new advances (either research or applications) or a combination of both.

The series emphasizes the following four areas:

**Mathematical Programming:** Including linear programming, integer programming, nonlinear programming, interior point methods, game theory, network optimization models, combinatorics, equilibrium programming, complementarity theory, multiobjective optimization, dynamic programming, stochastic programming, complexity theory, etc.

**Applied Probability:** Including queuing theory, simulation, renewal theory, Brownian motion and diffusion processes, decision analysis, Markov decision processes, reliability theory, forecasting, other stochastic processes motivated by applications, etc.

**Production and Operations Management:** Including inventory theory, production scheduling, capacity planning, facility location, supply chain management, distribution systems, materials requirements planning, just-in-time systems, flexible manufacturing systems, design of production lines, logistical planning, strategic issues, etc.

**Applications of Operations Research and Management Science:** Including telecommunications, health care, capital budgeting and finance, economics, marketing, public policy, military operations research, humanitarian relief and disaster mitigation, service operations, transportation systems, etc.

This book series is indexed in Scopus.

Fouad El Ouardighi • Gustav Feichtinger
Editors

# The Unaffordable Price of Static Decision-making Models

## Challenges in Economics and Management Science

Springer

*Editors*
Fouad El Ouardighi
ESSEC Business School
Cergy Pontoise Cedex, France

Gustav Feichtinger
Technical University of Vienna
Vienna, Austria

ISSN 0884-8289 ISSN 2214-7934 (electronic)
International Series in Operations Research & Management Science
ISBN 978-3-031-88637-9 ISBN 978-3-031-88638-6 (eBook)
https://doi.org/10.1007/978-3-031-88638-6



This Springer imprint is published by the registered company Springer Nature Switzerland AG
The registered company address is: Gewerbestrasse 11, 6330 Cham, Switzerland

# Foreword

Even though each of the eleven chapters of this book was independently written, they all contribute to a common theme and the book delivers a common message, which is that dynamic models and analysis are more appropriate for economics and management sciences than static ones, the latter seemingly being more prevalent in these fields. The editors of the book, Fouad El Ouardighi and Gustav Feichtinger, make a strong case for delivery of this position—what I would consider to be an irrefutable fact—by their selection of the individual entries comprising the book, preceded by their introductory chapter "Turning the page," discussing the overall goal and contents of the chapters. I say that superiority of dynamic models and analysis over static ones is an irrefutable fact, because of the richness such a framework provides, accounting for the most common type of decision-making process taking place over a horizon (and not be restricted to a single point in time), which is most natural for problems in economics and management sciences. This is the case regardless of whether there is a single decision-maker in full control of the evolution of the process (in which case optimal control theory provides the right set of tools), or even if there is a single decision-maker who however is not in full control of the process (and the outcome) because of dynamic (and stochastic) uncertainties or adversarial inputs (in which case robust optimal control theory or theory of zero-sum dynamic games provides the appropriate framework and the set of tools), or multiple decision-makers (players) strategically interacting under partially conflicting objectives (which calls for the theory of non-cooperative nonzero-sum dynamic games, with a plethora of information structures and its rich set of equilibrium solution concepts). This richness provides the researcher with the flexibility to come up with (and design) models that better capture "reality," and not be confined to myopic decision making which overlooks the long-term (or even relatively short-term but multi-stage) benefits of giving serious consideration (optimum in a precise sense) to how current stage decisions impact performance at future stages. In addition to this trade-off between stagewise gains and long-term (multi-stage) benefits, there is also the importance of the delicate balance a decision policy has to maintain between performance-driven action(s) and information to be transmitted to future stages for more accurate decision making in the future, which would result in better overall performance—two roles that would sometimes be conflicting, particularly in game situations with strategic policy making. Such issues of trade-offs and conflicts (known as *dual* or *triple* roles of decision policies), which are quite common in multi-stage decision making, necessitating also bringing in of a *learning* element, clearly do not arise in static/myopic decision making.

Many prominent economists, including several Nobel Laureates, have actually advocated dynamic models for economic decision making. For example, in their book *Robustness* (Princeton University Press, 2008), the two Nobel Laureates Lars Peter Hansen and Thomas J. Sargent work with stochastic dynamic models to address economic decision making when the decision-maker is faced with mis-specified models or does not fully trust the model at hand; see my review of the book in *Automatica* (95:511–513, September 2018). Two pillars of the approach advocated by Hansen and Sargent, as also acknowledged in the Preface of their book, are the connection established in the book (T. Başar and P. Bernhard, *H-infinity optimal control and related minimax design problems: A dynamic games approach*, Birkhauser, 1995) between robust control and a class of parameterized zero-sum dynamic games, and the approach to robustness in stochastic control problems through optimization of exponentiated cost (known also as risk-sensitive optimal control) as expounded in (P. Whittle, *Risk-sensitive optimal control*, Wiley, 1990), where the latter is also connected to parameterized zero-sum dynamic games (this time stochastic). This is of course all dynamic modeling and analysis, and to quote a statement made by the authors on pages 19–20 of *Robustness*: "*The 1950s–1960s control and estimation theories have contributed enormously to the task of constructing dynamic equilibrium models in macroeconomics and other areas of applied economic dynamics. We expect that the robust control theories will also bring many benefits that we cannot anticipate.*" This is a clear corroboration of the dictum that dynamic models are more appropriate (than static ones) for economic decision making—a position I have also advocated going back multiple decades, as in, for example (S.J. Turnovsky, T. Başar, and V. D'Orey, *Dynamic strategic monetary policies and coordination of interdependent economies*, The American Economic Review, 78(3):341–361, 1988; T. Başar and M. Salmon, *Credibility and the value of information transmission in a model of monetary policy and inflation*, Journal of Economic Dynamics and Control, 14:97–116, 1990). Among many others, the most recent (2024) Nobel Laureate, Daron Acemoglu, has also advocated the use of dynamic models for economic decision making, and he has predominantly used frameworks of stochastic control and stochastic dynamic games with also elements of learning in his particularly more recent work.

In concluding, I applaud the editors of this book for bringing the issue of decision making using dynamic models to the broader community through the contributions they have collected and put together as a coherent volume.

University of Illinois Urbana-Champaign Tamer Başar
Champaign, IL, USA

# Preface

At the *15th Viennese Workshop on Optimal Control and Dynamic Games* that took place in Austria in July 2022, many colleagues shared their perception that, though less effective than dynamic decision-making models, static models were increasingly widespread in several research areas of economics and management sciences, in terms of publications in scientific journals, attraction of doctoral students, fundraising for research, etc. This perception led us to publish a book that highlights the economic and/or social cost of founding decisions on static models in various fields of economics and management, and thus shows the superiority of dynamic models and methods in these areas. The dual objective is to explain the inability of static decision-making models to provide accurate economic and managerial prescriptions, and to promote dynamic approaches in economics and management sciences. We thereby encourage teachers, researchers, students, editors of scientific journals, and decision-makers to move beyond static approaches in economics and management science whenever possible.

We express our gratitude to the Springer Nature editors for their enthusiastic reception in favor of our work project, especially to Ms. Jianlin Yan. We are also grateful to Ms. Sneha Arunagiri for her coordination efforts and to Ms. Karen Sherman for the English editing. The editors appreciate the financial support from the research center of ESSEC Business School (CERESSEC).

Cergy Pontoise Cedex, France — Fouad El Ouardighi
Vienna, Austria — Gustav Feichtinger

# Contents

# List of Contributors


**Emmanuelle Augeraud-Véron** Bordeaux School of Economics, Bordeaux, France

**Hassan Benchekroun** Department of Economics, McGill University, Montreal, QC, Canada

**Alain Bensoussan** International Center for Risk and Decision Analysis, Jindal School of Management, The University of Texas at Dallas, Dallas, TX, USA

**Raouf Boucekkine** Aix-Marseille School of Economics, Aix-Marseille University and CNRS, Marseille, France

**Alessandra Buratto** Department of Mathematics "Tullio Levi-Civita", University of Padova, Padova, Italy

**Amrita Ray-Chaudhuri** Department of Economics, University of Winnipeg, Winnipeg, MB, Canada

TILEC, Tilburg University, Tilburg, The Netherlands

**Luca Colombo** Rennes School of Business, Rennes, France

**Christian Van Delft** HEC, Jouy-en-Josas, France

**Rodolphe Desbordes** SKEMA Business School – Université Côte d'Azur, Paris, France

**Gustav Feichtinger** Technical University of Vienna, Vienna, Austria

**Michael Freiberger** International Institute for Applied Systems Analysis (IIASA), Laxenburg, Austria

Wittgenstein Center (IIASA, OeAW/VID, University of Vienna), Wien, Austria

**Luca Grosset** Department of Mathematics "Tullio Levi-Civita", University of Padova, Padova, Italy

**Konstantin Kogan** Department of Management, Bar-Ilan University, Ramat-Gan, Israel

**Michael Kuhn** International Institute for Applied Systems Analysis (IIASA), Laxenburg, Austria

Wittgenstein Center (IIASA, OeAW/VID, University of Vienna), Wien, Austria

Vienna Institute of Demography, Wien, Austria

**Luca Lambertini** University of Bologna, Bologna, Italy

**Paola Labrecciosa** ESSCA School of Management (Paris Campus), Boulogne-Billancourt, France

**Andrea Mantovani** Department of Information, Operations and Management Sciences, TBS Business School, Toulouse, France

**Maddalena Muttoni** Department of Mathematics "Tullio Levi-Civita", University of Padova, Padova, Italy

**Fouad El Ouardighi** ESSEC Business School, Cergy, France

**Alexia Prskawetz** Research Group Economics, Institute of Statistics and Mathematical Methods in Economics, Vienna University of Technology, Vienna, Austria

Vienna Institute of Demography (VID), Austrian Academy of Sciences, Wien, Austria

International Institute for Applied Systems Analysis (IIASA), Laxenburg, Austria

Wittgenstein Center (IIASA, OeAW/VID, University of Vienna), Wien, Austria

**Gilles Rotillon** Paris-Nanterre University, Nanterre, France

**Miguel Sanchez-Romero** Research Group Economics Institute of Statistics and Mathematical Methods in Economics, Vienna University of Technology, Vienna, Austria

Vienna Institute of Demography (VID), Austrian Academy of Sciences, Wien, Austria

International Institute for Applied Systems Analysis (IIASA), Laxenburg, Austria

Wittgenstein Center (IIASA, OeAW/VID, University of Vienna), Wien, Austria

**Suresh P. Sethi** The University of Texas at Dallas, Richardson, TX, USA

**Dmitry Tsadikovich** Department of Management, Bar-Ilan University, Ramat-Gan, Israel

**Bruno Viscolani** Department of Mathematics "Tullio Levi-Civita", University of Padova, Padova, Italy

**Stefan Wrzaczek** International Institute for Applied Systems Analysis (IIASA), Laxenburg, Austria

Wittgenstein Center (IIASA, OeAW/VID, University of Vienna), Wien, Austria

# Chapter 1
# Turning the Page

**Fouad El Ouardighi and Gustav Feichtinger**

*Scientific knowledge is always the reformation of an illusion*
*Gaston Bachelard, Etudes, 1970*

**Abstract** In economics and management science, static models are not conceived as an intermediate step, a milestone, in the process of scientific understanding of a given phenomenon, but have become an end in itself. This chapter seeks to encourage teachers, researchers, students, publishers, and research fund providers to turn the page on static approaches in order to carry out research that is mindful of the future, more responsible research.

**Keywords** Static decision-making models · Myopia · Farsightedness · Responsible research

In a remarkable effort to establish a clear distinction between statics and dynamics in economic analysis, the famous economist Fritz Machlup (1959) reached the troubled conclusion that the division of economic analysis into statics and dynamics makes *too many* senses rather than *no* sense. Therefore, he advocated against the use of the terms statics and dynamics, and recommended that the terms be replaced by more meaningful concepts whenever possible.[1] At a time when economic analysis

[1] Machlup (1959) drolly summarized the cacophony reflecting the then prevailing controversial definitions of the two terms: "For more than twenty years I have been telling my students that one of the widespread uses of 'Statics' and 'Dynamics' was to distinguish a writer's own work from

F. El Ouardighi (✉)
ESSEC Business School, Cergy, France
e-mail: elouardighi@essec.edu

G. Feichtinger
Technical University of Vienna, Vienna, Austria



F. El Ouardighi, G. Feichtinger (eds.), *The Unaffordable Price of Static Decision-making Models*, International Series in Operations Research & Management Science 365, https://doi.org/10.1007/978-3-031-88638-6_1

was seeking to establish the main laws (or uniformities) of economics (Pareto, 1906), according to a partially prevalent, mechanics-based, consensus—mainly inherited from John Stuart Mill (1848)—the term statics referred to steady-state equilibrium characterization whereas dynamics referred to the transient path toward steady-state equilibrium analysis (e.g., Clark, 1898; Edgeworth, 1925; Hicks, 1939; Stigler, 1947).[2] Because both transient and steady-state analysis are inseparably complementary approaches in the mathematical analysis of dynamical systems, Machlup's conclusion of separating the proponents of the two approaches from one another may have seemed reasonable. In the current times, however, it appears somewhat indulgent. The reasons are threefold. First, economic analysis nowadays is mainly decision-making driven. In this regard, normative rather than descriptive models, involving one or several decision-makers, are sought in the distinct areas of economics and management sciences. Depending on whether a static or dynamic approach is adopted, such models may prescribe quite different courses of actions with potentially considerable stakes at hand. Therefore, a clear-cut, non-partisan position regarding the accuracy of the two approaches is needed. Second, whereas in Machlup's time, optimal control theory was still emerging, and no sufficiently established knowledge corpus on dynamic games was yet available, the progress of mathematical analysis of decision-making models during the last few decades has been considerable. This progress has expanded the set of methodological criteria of distinction between static and dynamic approaches toward decision-making. These criteria shed broader light to facilitate comparison between the two approaches. Finally, a marked preference for the static approach toward decision-making among researchers, scientific journals, and research funds providers can be observed regarding issues for which dynamics should naturally prevail. An obvious illustration is provided by the issue of environmental sustainability that has been tackled in the area of management science for two decades, most often with static models, from which managerial prescriptions should be derived for firms' use.[3] This is nonsense because this mainstream limits the corporate environmental responsibility to the level of instantaneous polluting emissions with no account of the effects of the pollution accumulation process. In this specific area, some top journals categorically reject research manuscripts dealing with the issue of environmental sustainability simply because they use dynamic modeling. In this context, PhD students are naturally discouraged from opting for the out-of-the-mainstream and thus risky path of dynamic modeling. Accordingly, funds in this area remain unduly invested in the less risky, yet potentially misleading, static model-based research mainstream. Therefore, static models are not conceived as

---

that of his opponents against whom he tried to argue. Typically, 'Statics' was what those benighted opponents have been writing; 'Dynamics' was one's own, vastly superior theory" (p. 100).

[2] Another partially prevalent consensus was instilled by Thorstein Veblen (1898) and, to some extent, also by Joseph Schumpeter (1943), who conceived of statics and dynamics as taxonomic and evolutionary approaches to economics, respectively.

[3] See the reviews by Kleindorfer et al. (2005), Corbett and Klassen (2006), Tang and Zhou (2012), and Jaehn (2016).

an intermediate step, a milestone, in the process of scientific understanding of a given phenomenon, but have become an end in itself. This phenomenon is all the more irrevocable as, because they are simpler to formulate and to resolve than their dynamic counterparts, these models can be learnedly and broadly shared in postgraduate and PhD programs, including with students who are the least enthusiastic about mathematical modeling.[4]

As noted by Ragnar Frisch in 1930, the distinction between statics and dynamics is a "distinction between two different ways of thinking, not a distinction between two different kinds of phenomena" (Frisch, 2011). Static decision-making models thus lead one to disregard the future consequences of current actions. That is, they promote myopic decision-making practices. In contrast, dynamic decision-making models seek to support farsightedness in decision-making. The virtues of farsightedness are obvious when one has to deal with decision problems involving stock variables besides flow variables (e.g., pollution stock besides instant polluting emissions, a firm's reputation goodwill besides its current advertising efforts, addictive behavior besides current consumption, a firm's inventory besides its current output, etc.). In fact, there are many more reasons to promote farsightedness in decision-making in economics and management science. We list a few of them below:

- First, a static decision-making model is unable to depict the whole spectrum of long-term possible outcomes associated with the resolution of a decision problem. In contrast, a dynamic decision-making model enables one to characterize the transient path of a system and the eventual complex behavior (e.g., limit cycling) associated with this path for any parameterization of the model.
- Relatedly, a static decision-making model prevents researchers from considering the possibility of history-dependent multiple solutions related to a decision-making problem, or the eventual case where such solutions are indifferent from the decision-maker's viewpoint. In contrast, a dynamic model can make such situations explicit and help to delimit indifference thresholds (usually called Skiba thresholds) related to transient paths and/or to the long-run outcomes, whenever these thresholds exist.
- Third, a static decision-making model makes it difficult to account for structural features that might affect a decision-making problem over time. Such structural features enable one to establish a categorization within a modeled stock variable, with heterogeneous reactions to specific actions. Dynamic decision-making models with distributed parameters are commonly used to disaggregate the modeled stock variable considered and improve the accuracy of the decisions inferred for each of the categories of the stock variable.

[4] In an effort to define a vision of responsible research in the area of operations management, Netessine (2021) stated several general principles, including the need for implementation of sound scientific methods and processes in both quantitative and qualitative or both theoretical and empirical domains. This effort will remain futile as long as farsightedness is excluded from the definition of responsible research.

- Fourth, a static decision-making model does not allow for the consideration of an abrupt change in the context of a decision-making problem over time. Such switching regimes can severely modify the set of feasible actions of a decision-maker. Dynamic decision-making models can easily include the possibility of a switch and enable decision-makers react accordingly.
- Finally, a static decision-making model does not take into account the structure of information in a multi-agent decision problem setting. This implies that such models cannot distinguish among the alternative modes of play that can be considered by non-cooperative decision-makers. In contrast, dynamic models can serve to characterize economic and managerial prescriptions that are contingent upon distinct modes of play, and thus provide a richer understanding of players' behavior.

Given the above explanations, one could reasonably paraphrase William Baumol (1968) to conclude that neglecting dynamic aspects in economics and management science is like playing Shakespeare's Hamlet without the Prince of Denmark. For our part, we would rather be inclined to consider the denial of dynamic approaches in these areas as the manifestation of scientific unawareness coupled with an inclination toward intellectual indolence. To some extent, "better complete ignorance than knowledge deprived of its fundamental principle," as professed by M. Gaston Bachelard (Bachelard, 1938).

Dynamic decision-making approaches have made considerable progress in terms of modeling and methods of resolution. Evidence is provided by the considerable number of widely acclaimed books published in recent decades, including those of Bensoussan et al. (1974), Tapiero (1977), Sethi and Thompson (1980), Başar and Olsder (1982), Bensoussan and Lions (1982), Feichtinger and Hartl (1986), Kamien and Schwartz (1981), Léonard and Long (1992), Dockner et al. (2000), Erickson (2003), Jørgensen and Zaccour (2004), Kogan and Tapiero (2007), Grass et al. (2008), Long (2010), Bensoussan et al. (2013), Kim (2017), Lambertini (2019), and Sethi (2022).

In view of such progress, it is important to make dynamic approaches more visible and accessible to the community of researchers, postgraduate and doctoral students, editors of scientific journals, policy makers, managers, research funders, etc., in economics and management science. The use of dynamic models should gradually become the rule, unless there are specific constraints that impose a static approach as an intermediate though self-degradable step. However, this requires active involvement by scientific journal editors in the areas of concern, to impose compliance with three major principles: diversity, fairness, homogeneity. First, editorial boards should be constituted in such a way that dynamic approaches are well represented in the journals of reference in each area. This represents a minimal requirement for ensuring methodological diversity in these journals. Second, the authors should be required to convincingly motivate the use of a static approach in lieu of a dynamic one, rather than the converse. By reversing the burden of proof against myopia, this would be a minimal condition for promoting methodological fairness among authors. Finally, only competent reviewers should be involved in the evaluation of dynamic approach-based manuscripts. This would be a minimal clause for ensuring homogenous methodological competency among reviewers.

This book is based on a widely shared belief that a shift from static to dynamic decision-making models might open up new opportunities for research in economics and management science. To this end, prominent experts in dynamic methods in both areas were invited to provide their particular perspective regarding the invalidity of static decision-making models in the form of very readable chapters. Below we briefly present their contents.

In Chap. 2, entitled "On the use of dynamic models in economics," Gilles Rotillon demonstrates the accuracy of dynamic modeling, though abstract and simplifying, for the analysis of the interaction between growth and the environment. The two most representative theoretical models related to this issue are presented, and their relevance with respect to the current societal debates is discussed. The chapter then states the relevant positioning of economists relative to the established social preferences and the pattern of consumption.

In Chap. 3, entitled "What to do with uncertainties?" Alain Bensoussan raises the issue of uncertainties. A number of important questions are considered: Can we improve our knowledge of these uncertainties? Is there a science of uncertainties? Is there an engineering of uncertainties? Do we have mathematical models of uncertainties? What is a product, if such a concept is possible for uncertainties? Understanding and mitigating uncertainties is an essential step in decision-making. This mitigation is related to obtaining information. Modeling and measuring information are key in the process, which is obviously dynamical.

Chapter 4, "Optimization in age-structured dynamic economic models," by Michael Freiberger, Michael Kuhn, Alexia Prskawetz, Miguel Sanchez-Romero, and Stefan Wrzacezk, presents the mathematical theory and potential applications of age-structured optimal control models. After presenting the general form of the problem and the related necessary optimality conditions, a model on air pollution is introduced, where consumption induces pollution, which in turn negatively affects utility, fertility, and mortality. The model is solved analytically and numerical simulations are then presented. The potential of age structure to solve non-standard optimal control models is finally demonstrated by considering optimal control models with random switches or time lags and delays.

Chapter 5, entitled "A vindication of open-loop equilibria in differential games," by Luca Lambertini, assesses the properties of open-loop equilibria in non-cooperative dynamic games, and illustrates the classes of such games that yield degenerate feedback strategies and equilibria under an open-loop information structure, and the resulting normative prescriptions.

In Chap. 6, "A linear-state game of advertising à la Vidale-Wolfe," by Luca Lambertini and Andrea Montovani, the tradition of advertising models stemming from Vidale and Wolfe (1957) is revisited to illustrate the possibility of building up a game delivering a (degenerate) feedback equilibrium under open-loop rules. To this end, assuming situations where advertising has an essentially predatory/defensive nature, the state equation of the generic firm is reformulated in such a way that its own advertising effort and the rivals' reaction to it enter the state dynamics additively. This modeling strategy results in a linear-state game structure, which can be resolved to provide an Arrovian result concerning the relationship between the aggregate advertising effort and industry structure.

In Chap. 7, "The cost of myopia with respect to a switching time in an advertising model, by Alessandra Buratto, Luca Grosset, Maddalena Muttoni, and Bruno Viscolani," the ability to react to abrupt changes is considered a fundamental skill for decision-makers, especially in dynamic contexts where problem structures can change over time. However, there are situations in which planners are myopic, i.e., unaware of the impending changeover, a context that inevitably results in a loss of profit. This chapter aims to assess the cost of adopting a myopic approach toward system changes in a marketing context. While the demand for a given product is influenced by the goodwill of the firm that produces, advertises, and sells it, the production costs may change abruptly with a hazard rate that depends on the demand for the product. An optimal control problem with stochastic switching time is thus formulated and resolved. Two situations are compared: The case of a planner who is aware of the possibility of a switch and that of a planner who is myopic with respect to such an event.

Chapter 8, "The limits of static decision-making rules in supply chain management, by Fouad El Ouardighi, Suresh P. Sethi, and Christian van Delft," shows that the use of static supply chain models can lead to wrong decisions. A series of simple issues representative of supply chain management are successively considered. For each issue, two versions of a supply chain game are defined, one static and the other dynamic. For the static version, an anticipative rather than a naïve formulation is adopted, wherein the repetition of the static game over a given time horizon accounts for the update of the previous period's considered performance on the current period. The decision rules and outcomes respectively inferred from the static and dynamic versions of the supply chain game considered are then compared. For each issue of interest, it is shown that the static decision rules provide distorted outcomes and misleading managerial prescriptions.

Chapter 9, "On the rebound effect of cleaner technologies and climate change: Radical technology innovations needed," by Hassan Benchekroun and Amrita Ray-Chaudhuri, starts by pointing out that technological innovations that reduce emissions per output can backfire and may result in countries increasing their emissions. In the case of climate change, assessing the size of this rebound effect requires a fully fledged dynamic analysis since the externality occurs across space and time. The welfare analysis needs to account for the sum of all generations' welfare. In a dynamic game, the impact of a technological innovation on emissions is ambiguous and depends on the initial stock of pollution. Therefore, relying on a simplified static version of the game or focusing the analysis on the steady state only can be misleading. Because the rebound effect may be strong enough to result in a decrease in welfare, it is advocated that policies aimed at fostering R&D in innovative clean technologies should target R&D projects with radical rather than incremental technological innovations.

Chapter 10, entitled "On the effects of an increase in the number of abaters in pollution abatement games," by Luca Colombo and Paola Labrecciosa, the effects of an increase in the number of abaters in pollution abatement games are investigated, first in a static and then in a dynamic (continuous-time) game. In both games, it is assumed that $m$ countries/agents agree on taking action to reduce the stock of pollution, which is a public bad, whereas $n$-$m$ countries free-ride on the abatement levels of the abaters. Moreover, abaters can either coordinate their contributions or not. In the static game, in both the coordination and the non-coordination scenarios, an increase in $m$ leads to a decrease in the stock of pollution and to an increase in social welfare. In the dynamic game, in contrast, in both the coordination and the non-coordination scenarios, an increase in $m$ may result in a higher steady-state stock of pollution and a lower social welfare, depending on the "business-as-usual" level of output. The authors conclude that the price of omitting the time dimension might be a wrong policy recommendation.

Chapter 11, entitled "Agroecology and biodiversity: A benchmark dynamic model," Emmanuelle Augeraud-Véron, Raouf Boucekkine, and Rodolphe Desbordes highlight how the choice between expanding agricultural land or retaining forest land is shaped by the bi-directional relationship between agriculture and biodiversity as well as the utility derived from biodiversity consumption. The static case shows that a high stock of biodiversity may be deliberately maintained as long as the agroecological productivity effect is important enough. This result also holds in the dynamic case. However, in the latter case, a large intertemporal discount rate can lead to total biodiversity loss along with the full collapse of the economy. It is also shown that the effect of a shift of consumer preferences toward agricultural goods (instead of biodiversity goods) on the biodiversity stock is much more ambiguous in the dynamic case than in the static case, depending on the strength of the agroecological productivity effect. These results have profound implications for biodiversity conservation.

Chapter 12, "Open-loop control-based linear-quadratic stochastic game with application to counter terror: Farsighted versus myopic policies," Konstantin Kogan and Dmitry Tsadikovich first notice that the typical solution to stochastic linear-quadratic problems in optimal control and differential game applications is based on feedback control. In contrast with real life, feedback control implies that the state dynamics are observable despite their stochastic nature. The authors overcome this unobservability drawback by deriving an open-loop equilibrium control for a linear-quadratic dynamic game with applications to counter-terror activities characterized by stochastic terrorist resource stocks. To do so, an open-loop Nash equilibrium solution based on expected terrorist resources rather than on the true state of the resource stock and its time-dependent feedback representation are derived. A comparison of the found equilibrium control with myopic behavior in response to resource dynamics by one or both parties shows that a farsighted party always has an advantage over a myopic party (i) under simultaneous commitments and (ii) when a farsighted party's leader openly commits to actions and the myopic party is a follower responding to the farsighted leader's actions. Furthermore, uncertainty improves the position of the farsighted party in terms of resource goals.

This book is intended to encourage teachers, researchers, students, and publishers to break with the established order imposed by the proponents of static approaches.

It is high time to turn the page on static approaches in order to carry out research that is mindful of the future, more responsible research.

## References


Bachelard, G. (1938). *La Formation de l'Esprit Scientifique. Contribution à une psychanalyse de la connaissance objective*. Librairie philosophique J. Vrin.

Başar, T., & Olsder, G.-J. (1982). *Dynamic noncooperative game theory*. Academic Press.

Baumol, W. J. (1968). Entrepreneurship in economic theory. *American Economic Review, 58*(2), 64–71.

Bensoussan, A., & Lions, J. L. (1982). *Applications of variational inequalities in stochastic control*. North-Holland.

Bensoussan, A., Hurst, E. G., Jr., & Näslund, H. (1974). *Management applications of modern control theory*. Elsevier.

Bensoussan, A., Yam, P., & Frehse, J. (2013). *Mean Field games and mean field type control theory*. Springer.

Clark, J. B. (1898). The future of economic theory. *The Quarterly Journal of Economics, 13*(1), 1–14.

Corbett, C. J., & Klassen, R. D. (2006). Extending the horizons: Environmental excellence as key to improving operations. *Manufacturing and Service Operations Management, 8*(1), 5–22.

Dockner, E. J., Jørgensen, S., Van Long, N., & Sorger, G. (2000). *Differential games in economics and management science*. Cambridge University Press.

Edgeworth, F. Y. (1925). *Papers relating to political economy*. Macmillan.

Erickson, G. (2003). *Dynamic models of advertising competition* (2nd ed.). Kluwer Academic Publishers.

Feichtinger, G., & Hartl, R. F. (1986). *Optimale Kontrolle ökonomischer Prozesse. Anwendungen des Maximumprinzips in den Wirtschaftswissenschaften*. Walter de Gruyter.

Frisch, R. (2011). General considerations on statics and dynamics in economics. In O. Bjerkholt & D. Qin (Eds.), *A dynamic approach to economic theory: Lectures by Ragnar Frisch at Yale University* (pp. 29–81). Routledge.

Grass, D., Caulkins, J. P., Feichtinger, G., Tragler, G., & Behrens, D. A. (2008). *Optimal control of nonlinear processes with applications in drugs, corruption, and terror*. Springer.

Hicks, J. R. (1939). *Value and capital*. Oxford University Press.

Jaehn, F. (2016). Sustainable operations. *European Journal of Operational Research, 253*(2), 243–264.

Jørgensen, S., & Zaccour, G. (2004). *Differential games in marketing*. Kluwer Academic Publishers.

Kamien, M., & Schwartz, N. (1981). *Dynamic optimization – The calculus of variations and optimal control in economics and management*. North-Holland.

Kim, B. (2017). *Optimal control applications for operations strategy*. Springer.

Kleindorfer, P. R., Singhal, K., & van Wassenhove, L. N. (2005). Sustainable operations management. *Production and Operations Management, 14*(4), 482–492.

Kogan, K., & Tapiero, C. S. (2007). *Supply chain games: Risk analysis and operations management*. Springer.

Lambertini, L. (2019). *Differential games in industrial economics*. Cambridge University Press.

Léonard, D., & Van Long, N. (1992). *Optimal control theory and static optimization in economics*. Cambridge University Press.

Machlup, F. (1959). Statics and dynamics: Kaleidoscopic words. *The Southern Economic Journal, XXVI*(2), 91–110.

Mill, J. S. (1848). *The principles of political economy*. John W. Parker.

Netessine, S. (2021). A vision of responsible research in operations management. *Manufacturing and Service Operations Management, 24*(6), 2799–2808.
Pareto, V. (1906). *Manuale di Economia Politica* (Vol. 13). Societa Editrice.
Schumpeter, J. A. (1943). *Capitalism, socialism, and democracy*. George Allen and Unwin.
Sethi, S. P. (2022). *Optimal control theory – Applications to management science and economics* (4th ed.). Springer.
Sethi, S. P., & Thompson, G. L. (1980). *Optimal control theory: Applications to management science and economics*. Springer.
Stigler, G. J. (1947). *The theory of price*. Macmillan.
Tang, C. S., & Zhou, S. (2012). Research advances in environmentally and socially sustainable operations. *European Journal of Operational Research, 223*(3), 585–594.
Tapiero, C. S. (1977). *Managerial planning: An optimum and stochastic control approach*. Routledge. 2 volumes.
Van Long, N. (2010). *A survey of dynamic games in economics*. World Scientific.
Veblen, T. B. (1898). Why is economics not an evolutionary science? *The Quarterly Journal of Economics, 12*(4), 373–397.

# Chapter 2
# On the Use of Dynamic Models in Economics

**Gilles Rotillon**

**Abstract** Our purpose is to show that dynamic economic modeling, far from being useless, has on the contrary the advantage to emphasize the main point, on the condition that results are not misinterpreted. To illustrate this point of view we analyze two canonical models of exhaustible resources of exploitation.

**Keywords** Dynamic modeling · Sustainable development · Exhaustible resources

## 2.1 Introduction

Economists often receive criticism for being capable of explaining why they were wrong yesterday. Theoretical economics, an esoteric "science" that escapes the real world in mathematical abstraction, is particularly targeted. The recent financial crisis that began in 2008 highlighted the risks of uncontrolled use of sophisticated models used to evaluate new financial products. Their complexity reserved them for the few "mathematical geniuses" who had designed them and they served as a reference for everyone to justify the valuations they predicted, thereby maintaining the belief in their reliability until reality suddenly reminded us that "trees don't grow to the sky." The bursting of the financial bubble has, in turn, raised suspicion about economic modeling in general, rekindling the debate about the autism of economists lost in their enchanted world(s) where equations replace real people as they act. This suspicion is all the stronger when we focus on the future with the ambition, if not to predict it completely, at least to shed some light on the directions it could

Reprint for Gilles Rotillon. Du bon usage des modèles théoriques en économie. *Revue Economique*, 66, 449–458, 2015. https://shs.cairn.info/revue-economique-2015-2-page-449?lang=fr

G. Rotillon (✉)
Paris-Nanterre University, Nanterre, France



F. El Ouardighi, G. Feichtinger (eds.), *The Unaffordable Price of Static Decision-making Models*, International Series in Operations Research & Management Science 365, https://doi.org/10.1007/978-3-031-88638-6_2

take. If the future time is one of the dimensions of the problem we are trying to understand, any attempt at modeling can only appeal to dynamic methods such as dynamic programming or optimal control, thereby taking the risk of being accused of de-realizing abstraction.

This text does not aim to continue this debate in all its generality, but more modestly to participate in it by showing that dynamic modeling, however abstract and simplifying it may be, can still be useful, provided that we do not expect more from it than it can give. In doing so, theoretical economists are neither "rational fools" à la Sen nor all-powerful demigods reading the future and dictating what should be in the name of Science, but participants in a public debate that they are not meant to close.

To illustrate this point of view, we focus on the relationships between growth and the environment as they have been studied over the past 30 years in academic journals along with the research on sustainable development. The first section presents the two theoretical models that seem to us to synthesize the essential on this subject. The second section is devoted to the lessons that these models teach us and their relevance to the corresponding societal debates. In conclusion, we return to the more general debate that we posed in the introduction and the position that the economist should, in our opinion, adopt.

## 2.2 Thirty-Five Years of Modeling Between Growth and the Environment

The aim here is not to provide a history of the modeling of the links between growth and the environment, nor even to present the most representative models. We want to stick to the essentials, namely the most general representations possible aimed at answering a key question: in what sense can we speak of sustainable development, given the appearance of new environmental constraints? The very idea of sustainable development, no matter what the precise meaning is, refers to the long term, making any attempt at modeling necessarily dynamic.

To pose it in all its acuity, the most standard cake-eating model below questions the intergenerational sharing of a depletable resource that can be considered "optimal." Optimal here means maximizing a criterion that is a formalization of a society's social preferences, which are characterized by its properties that we'll discuss further below.

$$\text{Max} \int_0^{\infty} U\left(c_t\right) e^{-\delta t} dt$$

$$\frac{dS}{dt} = -c(t)$$

$$c(t) \geq 0,\ S(0) = S_0$$

This depletable resource, whose initial stock is $S_0$, produces for society, represented by the function $U(.)$, a "utility" linked at each instant to the quantity extracted $c_t$. We put utility in quotes to indicate the high level of abstraction we are operating at. To fix ideas, one can think of a stock of oil whose flows serve a certain number of uses which are summarized here in the function $U(.)$. It is obviously possible to open the black box that is $U(.)$ and place a productive device inside, with several sectors, other resources, substitutability and complementarity, strategic behavior of economic actors, uncertainty, etc. All of this has been done, but the general message has not been changed, and it is this that we wish to emphasize.

Obviously, everything depends on the assumptions made about the utility function, i.e., the social preferences it represents. These (whose meaning we discuss in the second section) are usually as follows: $U(.)$ is increasing, is strictly concave, and satisfies the Inada conditions, $U'(0) = +\infty$ and $U'(\infty) = 0$. Under these assumptions, the optimal solution is to asymptotically exhaust the resource in such a way that the implicit price of the resource (equal to the marginal utility of the resource $U'(c_t)$) is constant in present value: this is the Hotelling rule. If the discount rate delta is zero, this problem has no solution.

Note the importance of the condition $U'(0) = +\infty$. This is what excludes the corner solution $c_t = 0$ for certain dates.

The problem with this model, whose message will be explained in the second section, is that it gives no importance to the resource as such. Only its use counts. Representative of the way our societies thought about their relationship to natural resources before the early 1970s, when the issue of the depletability of certain resources was not perceived as a problem, it is no longer sufficient to take into account the emergence of environmental issues and the values attached to them, regardless of the uses that are made of them. Today, nature is not (or no longer) just a reservoir of resources useful to humans, it increasingly has value in itself, and the question of its preservation (in a form to be determined) is urgent. Krautkraemer (1985) was the first to model this idea by introducing a utility function that depends on both the flow that can be obtained from the resource and its stock. In this case, the previous model becomes:

$$\text{Max} \int_0^{\infty} U\,(c_t, S_t)\, e^{-\delta t} dt$$

$$\frac{dS}{dt} = -c(t)$$

$$c(t) \geq 0,\, S(t) \geq 0,\, S(0) = S_0$$

This formalization in the spirit of Krautkraemer takes into account the amenity values provided by the environment (such as climate regulation by forests, the insurance functions implied by biodiversity, or the life-support functions enabled by certain environmental assets), but as Heal (1998) notes, these values make the environment a means rather than an end; however, environmental assets also have intrinsic value, regardless of their instrumental values. This is what monetary valuation methods used in environmental economics conceptualize as an existence value. At the level of abstraction where we stand, it ultimately matters little whether the formalization adopted, introducing the resource stock in the utility function, is interpreted as an amenity value or an existence value. What matters is that we recognize that the use of the environment as input alone is no longer sufficient to provide social utility and that a new trade-off between the direct use of the environment and its preservation appears. This formalization, undoubtedly simplistic, nevertheless emphasizes the rise of environmental issues in societal issues that have been observed in the last 40 years. These lead to the emergence of an environmental law, in the rise of ecological parties in political spheres, or the emergence of environmental NGOs and increasingly assertive opinion movements.

We can observe this new trade-off, for example, in the opposition between the supporters of the Notre-Dame-des-Landes airport (who prioritize $c_t$) and those who emphasize the preservation of the wetland that would disappear if the airport were built (thus wishing to maintain $S(t)$ at a certain level). However, if we maintain the Inada condition $U'(0) = +\infty$, it is clear that "consumption" $c_t$ (by "consumption" we mean here the use that society makes of the resource, whatever it may be) will always be positive and that introducing an environmental concern into the social representation does not modify the previous solution of asymptotically exhausting the resource. Therefore, we must assume $U'(0) < +\infty$.

In the case where the utility function is separable, that is, where $U(c_t, S_t) = U_1(c_t) + U_2(S_t)$, the optimal solution consists of reaching a steady state $(c^*, S^*)$ where $c^* = 0$ and $S^*$ satisfies $U'_1(0) = U_2(S^*)/\delta$. In other words, at the steady state, the marginal utility of consumption must be equal to the discounted marginal utility of the stock. Indeed, in the trade-off between consumption and preservation of the stock, the loss due to foregoing consumption is then exactly offset by the gain provided by the permanent increase in the stock and, given the assumptions on the utility function, $S^*$ is strictly positive. The proof of this result and the argumentation that validates it for a non-separable utility function can be found in Heal (1998). The interest of these abstract models is precisely to focus only on the essential: is a society that derives "utility" from a depletable resource sustainable? The society is "simply" represented by $U(.)$ with the properties attributed to it, and the only resource it can use is depletable. It is obviously a much less favorable case than if it were renewable, and has the added advantage of representing one of the essential features of our mode of production and consumption. This fundamentally relies on the extraction of many depletable resources, foremost among which is oil. With a renewable resource, it is obviously possible to achieve balanced growth while maintaining the resource stock constant, since it is only necessary not to extract more than its own reproductive capacity. Certainly, in practice, we use renewable resources too intensively, transforming them, alas, increasingly into depletable resources. But this is less an economic problem than a political problem of poor management of the stocks of these resources. The real constraint on development that can be described as sustainable in the sense that, as Solow (1993) indicates, "something is conserved over the very long term" is the intensive use we make of depletable resources, foremost among which are fossil energy resources. Moreover, introducing a productive activity where the resource is no longer directly the source of social utility but a simple factor of production adds nothing essential but only places the problem at the level of the means to be used to respond to social preferences. These are certainly important issues, which have been the subject of most of the academic literature of the last 40 years, where the more or less strong substitutability of factors of production, the role of technical progress, externalities, etc., are fundamental. But these are still secondary issues if the answer to the previous question about the existence of sustainable development is negative, and it is on this question that we intend to focus. It is now time to leave the sky of theory and return to the more solid ground of interpretations.

## 2.3 What Are the Lessons to be Learned from These Models?

What this first model tells us is that a society that values only the use it makes of a depletable resource can only deplete it, extracting less and less as it becomes exhausted if it has a preference for the present. We can ignore the case where this preference for the present would not exist ($\delta = 0$) on the grounds that such a society does not exist in the present world. Thus, in such a society, sustainable development is not possible, regardless of the sense given to this term, which can only concern utility (constant $U$), the resource (constant $S$), or its use (constant $c$). Furthermore, such a society is not intergenerationally equitable since it favors present generations over future generations, even though social preferences (the "utility" function) remain constant. One might think that such a society would not be socially sustainable in the sense of a social contract that could be accepted by all. One might also think that this model illustrates quite well the functioning of developed societies from the early twentieth century to the early 1970s, where environmental constraints arising from the existence of depletable resources, such as oil, were ignored, and that it highlights the reason for this mode of operation and the impasse to which it leads, namely collective preferences for "consumption," i.e., the use made of the resource. This "preference for consumption" is at the heart of assumptions about the utility function. Its increasing nature means non-satiation, with greater consumption always implying greater social utility. Similarly, the condition of Inada $U'(0)$ equals to infinity implies that consumption is always strictly positive, and it is this assumption that necessarily leads to the asymptotic depletion of the resource. In our modern consumer societies, these two assumptions fairly represent the consumer who can never really be satisfied since he would then have no further demand. He is therefore condemned to a runaway consumption, supported by advertising that aims to transform desires, which are by nature infinite and never satisfied, into needs. If such a society is built on the exploitation of depletable resources, this first model tells us that it can only deplete its resources, thus endangering its very existence. The second model confirms this result, since the introduction of an environmental "concern" (in the sense that the environment provides social utility independent of its uses, and expressed in the formalization by the introduction of the stock in the utility function) does not modify the previous result if the "absolute preference" for consumption implied by the condition $U'(0,.) < +\infty$ is not modified. Hence the first result provided by the second model: it is necessary to modify social preferences regarding the use of depletable resources if sustainable development is sought. At this level of abstraction, this formalization can be accepted as much by a convinced utilitarian as by a militant ecocentrist. This conclusion is also reinforced by that part of the literature that questions the criterion that formalizes social preferences. These developments, which were initiated by Chichilnisky (1996) with her axiomatic reflection on sustainable development, show clearly that optimal trajectories are modified in the direction of greater conservation of the resource when the social choice criterion explicitly takes the very long term into account. In this second model, it is possible to reach a stationary state (and therefore sustainable) where unused resources will be preserved. However, we may be surprised by the "mathematical" result and return to the economists' interpretation of autism that it could illustrate. Because what this model strictly tells us is that we will reach a situation in finite time where we will consume nothing and where utility will only come from contemplating the unused resource ($S^*$ strictly positive). The love of nature (and mathematics) leads straight to decay! Even advocates of degrowth do not go that far.

This is forgetting the abstract nature of the model and taking c(t) for real consumption and not for its concept. Just as Spinoza taught us that the concept of a dog does not bark, the concept of consumption does not feed. What this model very concretely tells us is that what we called "consumption" in the previous model is no longer essential to provide social utility. In other words, sustainable development based on exhaustible resources is only possible if our consumption patterns change. This precisely reflects one of the societal themes at the heart of questions about the possibility of sustainable development, and which everyone knows more or less subconsciously is not possible by hoping to generalize the American (or even Italian or Portuguese) way of life to the planet. The link between social preferences and lifestyle needs to be explained. In these models, the former refers to the properties of the utility function and is, as usual in this formalized approach, primitive data, while the latter refers to the control variable c(t) which reflects the pressure that preferences exert on the environment. It follows that since preferences are primitive, they generate the "optimal" depletion of resources and thus lead to the resulting "way of life." But in practice, awareness is most often raised by the realization that our lifestyles are not sustainable. To take just one example, this is what the successive reports of the IPCC continue to illustrate. And if the world's current population had a final energy consumption level equivalent to that of Americans (an average American currently consumes 4.9 toe/year compared to 0.5 for an African), global consumption would be triple the current level and 50% more than the global consumption projected for 2050 by the IEA.

We must emphasize the significance of this interpretation, as it is not very intuitive for many economists who, in our view, remain too confined to their theoretical models. During a seminar where we presented this work, we were criticized, on the one hand, for presenting an unrealistic result since zero consumption is contrary to the very existence of any life and, on the other hand, for having an outdated interpretation since it does not take into account the introduction of endogenous technical progress à la Acemoglu (2009) allowing us to compensate for resource depletion. On the first point, it should be emphasized that the fact that consumption is zero at the stationary state is not our result but the classical one that can be found in the literature and was perfectly explained by Heal (1998) without anyone raising any criticism. Our "result" lies in the interpretation we give to this impossible zero consumption and seeks to give it a meaning which it seems a priori devoid of. On the second point, it should be noted that endogenous technical progress à la Acemoglu is at the very least ad hoc with a continuum of firms all doing the same thing at equilibrium! If we criticize zero consumption on the grounds of its unreality, we cannot defend such a conception of technical progress, even if it is called endogenous.

Finally, we can see that these models, with such a high level of abstraction that they may seem irrelevant, actually highlight that the two main questions that should be at the center of public debate on sustainable development are those of social preferences (translated in the language of modeling by the properties of the utility function) and the lifestyle that we must adopt (synthesized here by the variable $c(t)$). Of course, then come the questions of the means to use to achieve the goals that we have set ourselves. But if we do not realize that this excess consumption precisely comes from our social preferences, the implementation of means aimed at changing this lifestyle without changing these preferences is necessarily doomed to failure. However, when we look at the public debate on these issues, we rather find an absence of questioning of our lifestyles than a reflection on defining new, more sustainable ones. On the other hand, there is an abundance of proposals on the instruments, technologies, and institutions that should be implemented to continue as long as possible on the trajectory that we are still following for the moment, and that our simple models tell us is not sustainable. Thus, in a very concrete debate on such regulations (such as a carbon tax), such protocols (such as those from the various COPs that remain ineffective), or such technology presented as a substitute for fossil energy (renewable resources accounting for 6.7% of global energy consumption compared to 82% for fossil energy), we should first question the role they play in redefining our social preferences and lifestyles. For example, we can analyze the debates around the aborted French carbon tax project as a debate between the majority of the social body that highlighted the problems of wealth transfers it would have involved and those who saw it as a necessary means of transforming behaviors to combat carbon emissions. The former remained in the model where consumption in its current form is a priority, while the latter sought a way out. It may be regrettable that, like Ulysses, we have to impose a constraint to escape the sirens' songs, and we can dream of a better solution, but given the current state of the problem and social preferences, a return to the status quo is clearly not the right choice since it implies a continuation of carbon pollution. And what our model tells us is that this choice is not sustainable in the long term. And this debate is much more fundamental than the academic debate where some are between so-called weak sustainability and strong sustainability. The former approach, initiated by Solow (1974), then by Hartwick (1977), considers that only the aggregate stock of capital counts, which allows the substitutability of different forms of capital (natural, human, produced), while for the latter, of which Daly (1977) is the precursor, natural capital is not always substitutable for other forms of capital. When it is the case, the sustainability of the economy requires to preserve these forms of "critical" capital above certain thresholds to be defined. A balanced presentation of the two approaches can be found in Neumayer (2003), but most of the time, they are laid out as antagonistic, as in Vivien (2005). Again, this is confusing the formal results of abstract models and their meaning. For example, emphasis must be set on technical progress just as the potential means to bring out this substitutability. And if there is opposition between the two approaches, it lies mainly in the ways that must be prioritized in the search for sustainable development, with technical progress for one and the definition of thresholds and critical capital for the other. This doesn't necessarily appear irreconcilable, but it positions this debate as secondary to the one underlying our two models.

## 2.4 Conclusion

One could argue that the above reflects a narrow conception of economics, with the two models presented appearing representative of the so-called neoclassical approach which, while undoubtedly dominant in universities, is increasingly being questioned by both "heterodox" economists and specialists from other social sciences. However, we are not here engaging in an epistemological debate on the validity of the neoclassical representation, which lies at the level of the assumptions made about preferences (complete pre-order, transitivity, continuity), but rather at a higher degree of abstraction that does not put any particular content a priori on what is meant by "social preferences" or "utility." On the contrary, the content of social preferences is precisely the subject of debate, and its definition, necessarily contingent on the social context, cannot be posited a priori. This means that here ends the economist's power.

What place do these models then give to the theoretical economist? Certainly not the central place that some might have hoped for, since the essential choices are not within his or her purview. The normative economist, advisor to the Prince from the heights of his or her knowledge, is undoubtedly challenged, and the 2008 crisis triggered by the subprime crisis should convince even the most stubborn among them. He or she does not have to define social preferences and, until recently, has not been particularly concerned about them, considering these preferences, against all evidence, as given and immutable. But he or she is not useless, if he or she can translate what his or her models tell him or her into clear language, and deliver them to the public so that they can use them as compasses that constantly bring us back to verify that we are on the right track. Of course, the compass is not enough, but without it, we are quickly lost. And ours indicates, first, that we must work to modify social preferences that currently favor the status quo, and second, to rethink our consumption patterns. It can be added that these two orientations are sufficiently general not to be reserved for specific actors (experts, companies, politicians, etc.), but rather concern all members of society, especially since it is not possible to impose specific social preferences, which is why we attach importance to public debate, which alone can lead to a modification in our behavior that is freely accepted in the end.

Finally, the above results should not be overestimated. Knowing that a model allows us to think that sustainable development is possible (in terms of the constancy of certain indicators, here consumption and resource stock) does not tell us how to achieve it in the real world. One thing is to know that we are heading in the right direction (i.e., discussing our consumption patterns, our social preferences, and thus the type of society we want); another thing is to choose the best means to achieve it. Here we return to the realm of concrete policies to be implemented, which were explicitly excluded from the abstract models we have discussed and which also require other informed public debates. It remains that, for the moment, the North that our compass points to is not the direction in which we are heading.

## References


Acemoglu, D. (2009). *Modern economic growth*. Princeton University Press.

Chichilnisky, G. (1996). An axiomatic approach to sustainable development. *Social Choice and Welfare, 13*(2), 219–248.

Daly, H. (1977). *Steady-state economics*. Earthscan.

Hartwick, J. (1977). Intergenerational equity and the investing of rents from exhaustible resources. *American Economic Review, 67*(5), 972–974.

Heal, G. (1998). *Valuing the future*. Columbia University Press.

Krautkraemer, J. (1985). Optimal growth, resource amenities and the preservation of natural environments. *Review of Economic Studies, 52*, 153–170.

Neumayer, E. (2003). *Weak versus strong sustainability* (2nd ed.). Edward Elgar.

Solow, R. (1974). Intergenerational equity and exhaustible resources. *Review of Economic Studies, Symposium* (pp. 29–46).

Solow, R. (1993). An almost practical step toward sustainability. *Resources Policy, 19*(3), 162–172.

Vivien, F.-D. (2005). *Le développement soutenable*. La Découverte.

# Chapter 3
# What to Do with Uncertainties?

**Alain Bensoussan**

**Abstract** The objective of this chapter is to discuss the issue of uncertainties. Uncertainties are certainly a reality, and we do not know them. Can we improve our knowledge of these uncertainties? Is there a science of uncertainties? Is there an engineering of uncertainties? Do we have mathematical models of uncertainties? What is a product, if such a concept is possible for uncertainties?

**Keywords** Uncertainty · Probability · Decision making · Risk management · Information

## 3.1 Introduction

Sciences like Physics, Chemistry, Biology, Economics, and others are concerned with understanding better and better the reality of our world. It is a common goal; they are simply specialized in sectors of this reality. This reality is not really known, but scientists are able to formulate theories or models of this reality. At any time, the job of scientists is to develop convincing arguments to assert that their models are close to reality and the closest possible at the current time.

Alongside sciences, we have Engineering with all possible subdomains. We have Medicine, we have Management Sciences and many others, which develop, from the progress of scientific knowledge, a multitude of applications which improve our living conditions. They use the models of sciences, extend, or adapt them, or develop specific ones. They can be called applied sciences, since the major difference with the fundamental sciences lies in the sector or domain of interest, which is more focused because products should be the outcome of the research.

A. Bensoussan (✉)
International Center for Risk and Decision Analysis, Jindal School of Management,
The University of Texas at Dallas, Dallas, TX, USA
e-mail: axb046100@utdallas.edu

 
F. El Ouardighi, G. Feichtinger (eds.), *The Unaffordable Price of Static Decision-making Models*, International Series in Operations Research & Management Science 365, https://doi.org/10.1007/978-3-031-88638-6_3

Mathematics is also called Mathematical sciences. However, they are different from the other ones. Mathematicians are not interested in a specific domain of reality. They are equally interested in all of them. If fact, they develop a language and techniques of reasoning, which appear to be the most efficient possible for the models used by the other scientists to improve their knowledge of reality. It is a remarkable fact indeed that the most useful models of reality are mathematical ones. Quite interestingly, it may appear that the same mathematical model can be adapted to two different sciences, for instance, physics and economics. Of course, computers are essential, mostly because they can implement mathematical models fast and speed up the progress toward improving knowledge and obtaining applications and products.

The objective of this paper is to discuss the issue of uncertainties. Uncertainties are certainly a reality, and we do not know them. Can we improve our knowledge of these uncertainties? Is there a science of uncertainties? Is there an engineering of uncertainties? Do we have mathematical models of uncertainties? What is a product, if such a concept is possible for uncertainties?

Understanding and mitigating uncertainties is an essential step in decision making.

This mitigation is related to obtaining information. Modeling and measuring information are key in the process. It is obviously a **dynamic** process.

## 3.2 Historical Vision

It is important to realize that the lack of knowledge in the real world was not considered in history as uncertainties. Religion was the reason. Since God has created the world, the fact we have a limited knowledge of its reality is simply due to our own limitations. It is quite legitimate to improve our knowledge. It does not make man equal to God, in doing that. But if uncertainties represent something which will occur in the future, how can we pretend to know what will happen? The decision of what will happen relies on God. How can we pretend knowing God's decisions? It will position Man at the level of God.

If, conversely, we consider that God does not interfere much, it is different. Of course, the world exists, and we constantly increase our knowledge about it, but we will never have a full knowledge. We will never know what happened before the Bing Bang. We can leave the creation to God, or accept that we will never know, and focus on the uncertainties of the future. For them, if God does not interfere, we may act and see what we can do about uncertainties. This explains why the scientific approach of uncertainties was considered much later of the scientific approach of Nature.

## 3.3 A Change of Paradigm

We can do a lot for uncertainties. We can apprehend the uncertainties in the future, first because they are largely consequences of what happened or was done in the past. Second, the source of uncertainty is, quite often, complexity. In the real world, many aspects interfere. Even though individual aspects can be understood, their combination can have intricate consequences which can be unpredictable.

In these two cases, mathematical modeling will help a lot. We will discuss this in the next section.

The engineering part of uncertainties is risk analysis and decision making. We cannot separate uncertainties from their impact. At the end of the day, it is the impact which is significant. Here again, mathematics will be essential.

In this set up, the science and the engineering of uncertainties exist and are comparable to other sciences and engineering applications.

## 3.4 Mathematical Concepts and Models

The main mathematical theory developed to deal with uncertainties is probability theory. As usual with mathematics, it offers a model of reality, not reality itself. This model can address any uncertainty, very simple ones as well as very sophisticated ones. For simple ones, where intuition can suggest results, it should coincide with what the intuition provides and, thanks to the power of mathematics, should also provide the solution in complex cases, where intuition is powerless.

For instance, if the uncertainty relates to the outcome of rolling dice, the intuition will say that there are 6 possibilities, which are equally likely, unless the dice is flawed. We immediately associate to each outcome a probability of 1/6. But if we are interested in the income of Americans, compared to the income of Chinese what can probability theory do for us?

We need to get some information and start reasoning. To simplify, we may split the wealth into tranches, for instance, from 0 to $20,000, from $20,000 to $40,000 up to $200,000 and one tranche for wealth above $200,000. We need to associate to each tranche a probability. This is the basic element of the construction of the model. Probabilities are decisions of the model builder, obtained from some information and some reasoning. Obtaining information is like realizing experiences in physics. Ideally, we should count the number of Americans in each tranche. Nobody will object that the probability of a tranche should be chosen as the ratio of the number in this tranche divided by the total number of Americans. This is called the frequency. Since acquiring the information in this way will be extremely costly, we proceed with taking a **sample** of say 1000 American. This sample should be representative of the population, as far as wealth is the characteristics of interest. For a sample, we can do what was mentioned above for the full population and choose as probabilities of tranches, the frequencies of tranches in the sample. We may try to improve this choice, by all techniques of learning. This is the job of mathematicians. We build the model of approximation when we decide that the frequencies in the tranches of population are the same as those of the sample and call them probabilities that the wealth belongs to the respective tranches.

The wealth of Americans is then considered as a **random variable**. This reflects the uncertainty, since it is not a deterministic number, and the various probabilities constitute a probability distribution.

A big breakthrough arose when probabilists agreed that probabilities could be considered as measures of uncertainties, in a way very similar to measures of length, surface, or volume in physics. This is particularly helpful to get rid of tranches and address a continuum of randomness. This analogy with measure was quite precious because mathematicians had developed a very sophisticated theory of measure, with many useful results, which could be immediately used in probability theory.

Consider next uncertainties due to the fact that we are interested in events in the future. This is clearly a very common situation. For that purpose, probabilists developed the theory of stochastic processes. It has led to an impressive number of concepts, methods, and results.

We mentioned above the information obtained by sampling. It is clear that in the case of stochastic processes, the issue of information is essential, since we are interested in predicting some future event, based on the information obtained in the past. Probabilists were able to quantify information, to measure it. After collecting all the progress made, we can rely now on very efficient tools to address complex uncertainties, including those which will occur in the future.

It is amazing to see how vague concepts like independence or dependence of information have been mathematically defined in a way that allows quantification even for complicated sets of information. At the same time, it is also important to recognize limits. For instance, the mathematical apparatus is only able to take account of information which increases with time. No possibility of forgetting is allowed, no possibility of degradation can be considered. This concerns the general mathematical theory of course. In practice, some treatment cannot be avoided.

Why is that so being not clear. It is part of the mystery of mathematics. If we insert a slight change in a model, which works very well, everything may collapse. This is true for Probability theory. It explains that it has been challenged as the right model for uncertainties. Data Analysis has been one attempt. We should learn from reading data, without any probabilistic set up. Another attempt has been Fuzzy Sets. Now we see Data Science. We can observe that probability has resisted all attempts to supersede it. It is the best compromise between a good level of approximation of reality and possibilities of reasoning with significant results.

An important approach of probabilistic models for practical purposes is simulation. It is the opposite of deriving probabilities from frequencies. For a given probability distribution we can generate a sample whose frequencies are close to the probabilities of the distribution. How it is done is magic. Indeed, algorithms used by computers are always deterministic. Mathematicians can design algorithms whose results look random. They fool us for a good cause.

Probability theory is adequate for what is called "Risk Analysis," namely, describing uncertainties. But Risk analysis is part of the story, it is not the end. The end of the story is Risk Management, which is the methodology of decision making under uncertainty. Once we have a probabilistic description of risks, how do we act?

The general idea is that of cost-benefit analysis, taking account of uncertainties. We may first try to reduce uncertainties, by acquiring more information. This is always costly, in terms of money and time. At some point, the additional information which can be obtained is not worth the cost. It means that at the end of the day, we must decide even though there remains uncertainty. How can mathematics help?

Let us go back to history and consider the situation after World War II, with the development of Operations Research, which originates in the difficulties of logistics related to military operations. The issue was efficiency. How to produce fast and abundantly, with limited resources. The mathematicians introduced optimization under constraints, scheduling, queueing theory, and many other developments. This was the core of Management Science from the war to the mid-seventies. We were helped by economic growth, and a stable political situation, based on nuclear deterrence.

Uncertainties were important in two basic cases, reliability and quality control in industry, and financial risks in insurance and banks.

Life Insurance is a good example of how stochastic optimization works. Age of death is a random variable, but its probability distribution is sufficiently known. To pay a premium each year for a sum to be given by the insurance company to the spouse of the insured makes a lot of sense. It must be computed so that the insurance company is profitable, but it should not be too high; an easy problem of stochastic optimization.

Financial markets offer another example of stochastic optimization for which mathematicians have provided many methods and tools. Assets are described as stochastic processes, whose evolution and correlations have been subject to a lot of research, in what is called Mathematical Finance. The optimization of investment decisions is a well-known problem, and a solid theory exists. It justifies the idea of diversification, which comforts a commonsense attitude.

Risk Management needs to introduce the impact of risks. What are the consequences of the failure of equipments, what are the consequences of investments in a financial product?

Once this is apprehended, mathematicians have worked a lot to create tools and methods to help decision makers.

Is the situation of decision making under uncertainties satisfactory, thanks to the progress of the past decades? The answer is unfortunately no, for reasons addressed in the next sections.

## 3.5 Big Changes in the Past 50 Years

From a world where risks and uncertainties were limited and well apprehended by mathematical methods, we have evolved to a world with new risks, which are huge, structural, and diverse. Can we use mathematics? Much less. Mathematics is adequate when risks are of technical origin, or natural. The major change is that the new risks we are facing now have been mostly created by **us**. Even when we consider innovation and technology, the fantastic development of AI has become the source of huge concern, because we do not use it rationally. Similarly, the Covid pandemic is very likely originated by human failures, and its management has been chaotic. The political situation has evolved from a stable confrontation between rational players to a multipolar world involving dictatorships and religious fundamentalism.

Climate Change is also related to Humans. It is, however, more rational. We realize that we have a common challenge, but the price to pay for a solution is naturally a matter of harsh discussions. Nevertheless, it is quantifiable and negotiable. At any rate, in the case of climate change, we may be in a situation where risks occur so often and periodically, that they can be considered as certain. Probability theory is not relevant if probabilities are very high. We are in a deterministic context. Think of storms and hurricanes in the US, as a good example.

We must accept that, within this framework of new risks, quantitative and mathematical techniques cannot apprehend all aspects. Opinions are not the consequence of rational reasoning based on facts, but the consequence of charismatic speeches on social media, made by extremists.

Leadership is needed to confront this new reality. It is important to understand that when leaders are failing, there is little hope to see subordinates acting rightly.

What makes a good leader remains largely an open question.

## 3.6 What to Do with New Uncertainties?

We focus on enterprises because for government one needs to incorporate aspects which are not present in corporations. A parallel analysis could be made for the government.

The evolution of uncertainties does not imply that those of the past have disappeared. An enterprise must still offer products and services and be profitable.

An enterprise must still innovate and adapt to opportunities. Technical risks remain present and all the quantitative techniques to handle them remain valid.

But Enterprise leaders must have comprehensive perspectives: profit is more a constraint than the main objective. The perspective is to understand the risks and to control them for all activities and decisions. If production is done abroad, having an efficient production tool is not the issue. The issue is to be sure of the reliability of suppliers.

Alongside technical aspects, the CEO should be aware of psychological aspects and organizational aspects.

## 3.7 Psychological Aspects

The fact that Humans are essential players in social systems is of course known for long. Mathematicians have worked on this issue and proposed methods which are part of Risk Management. Decision theory and Game theory are the main scientific outcomes of this effort. For instance, a clear psychological fact is that humans do not like to decide on probabilities. They want to be sure, or at least to decide on indicators of risks which are simple and adequate. In that regard, a huge methodology of risk indicators has been developed. They provide a successful approach to the problem, in particular for financial decisions.

Another well-known psychological aspect of decision making under uncertainty is risk-aversion, or its opposite risk-appetite. Can we have a quantitative approach? The theory of utility function has been developed successfully in that regard.

Unfortunately, mathematics can help in modeling human behavior only when there is rationality, or at least a clear understanding of what could look irrational.

It turns out that the biggest catastrophes have resulted from what is called the **taboo effect**, or what may be called the **blindness of Ego.** The bigger the risk, the less we want to consider it. To some extent, we are blind to things that others see quite well. It is very likely that Ego has a lot to do with this blindness, but Greed can also be a reason. For instance, why so many smart people have been gullible to Madoff? There is also an opposite effect: an excessive **precautionary principle.** The public emotion disseminated by media may compel governments to take hasty and unreasonable decisions. Nuclear energy is a good example. A reasonable approach would be to compare this source of energy to others, analyzing all consequences on climate change, on the stability and reliability of sources, on geopolitics and global economy.

People, Societies have not, by themselves, a rational attitude toward big risks. They are too emotional. They also can be too permissive. Why, hard to tell! In 2010, the offshore drilling rig Deepwater Horizon of British Petroleum exploded in the Gulf of Mexico, originating catastrophic damages. Nevertheless, in October 2011, BP got the green light from the US government to drill again in the gulf.

In projects of this kind, where public opinion is so irrational, the best is to be very transparent, with all stakeholders, including the public; to provide as soon as possible all relevant information on opportunities as well as on risks. It may be easier to say than to do. Indeed, it works if there is trust. After Fukushima, the Japanese had lost trust in their government, when they discovered that the regulating agencies were not really independent; a big difference with countries like the US or France, where regulatory agencies are fully independent, and trust exists.

## 3.8 Organizational Aspects

We can say with quasi-certitude that at the origin of the most catastrophic events there is a human failure, or at least a blatant lack of preparation. This is not only true for banks, where greed is often put forward. It is also true for highly technical domains like space, where greed cannot be the reason, at least within space agencies. It may look paradoxical for a domain like space, where standards of security are so high and where norms and procedures are examples for industries like aeronautics, railways, and cars. However, the two major accidents of the shuttle, Challenger in 1986 and Columbia in 2003, are due to human failures, which were predictable and should not have occurred. No wonder that it is also true for all industries and naturally the financial sector.

The reason is simple. Humans are governed by emotions more than by rationality. It can be greed, but more simply and universally ego. In addition, we forget the lessons of the past, even more because situations cannot be exactly similar.

For any institution, including naturally corporations, the only way to mitigate risks and to be ready if they occur is to put in place a solid organization, which involves all members, starting at the level of the CEO, who must be convinced of the need for efficient risk management. The CEO must promote the culture of risk across the company, explain how helpful it is and that it is not an additional bureaucratic layer.

Among the basic elements of culture, there is the idea that risk management must be comprehensive. Especially in view of the new risks, focusing on the technical risks related to the production process and neglecting external risks can be catastrophic.

Fortunately, many enterprises have made progress toward the creation of a strong risk management structure. A new position has appeared within the Board of Directors, that of CRO (Chief Risk Officer). The CRO is at the head of a risk management unit. Among the major tools, there is the ERM (Enterprise Risk Management) which is an information system integrating all elements of risk analysis and risk management. The goal of the structure and the related tools is to get a risk intelligent organization to obtain the best risk-informed decisions.

It is very important to clarify that the CRO is not responsible for the risks. Each operational unit bears the responsibility for its own risks. The CRO is responsible for providing the common framework and for the fact that nothing is forgotten and that lessons of the past have resulted in necessary changes. The CRO is responsible for building trust with all stakeholders, including employees and the public.

## 3.9 Framework of Methodology

Quantitative techniques are essential, since at the end of the day we need to quantify to compare risks, to allocate resources and take the best risk-informed decisions. An important difference concerns the case of a single event, for instance, the occurrence of an accident, versus a continuum of uncertainties, for instance, the analysis of future revenue or future income. In the case of a single event, the mathematical set up is limited. One needs to define the probability of occurrence and the impact (amount of damage) of the event. Of course, obtaining a precise assessment of these two numbers may be complex and may require some sophistication.

This is particularly true for systems. If we ask the question of probability of a crash of a plane flying tomorrow, we address a system, not a specific equipment. The calculation depends on the reliability of each part or each equipment and on the correlations between the failure of various parts. In the literature, risk analysis for systems is called PRA, probabilistic risk assessment.

Finally, one obtains a simple and efficient indicator of risk by taking the product of the two numbers. The comparison of two risks is easy with this indicator. In the case of a continuum of uncertainties things are more complex. The full power of probability theory and stochastic processes may be necessary if time is an element of the model. There is no immediate indicator in general. We rely on decision theory, on stochastic programming, and on stochastic control to obtain an optimal solution. When there is time in the problem, the issue of what is a decision comes in, because the information depends on time. The decision must depend on the level of information which is available when it is applied. What we look for is a decision rule, which tells us what to do in each circumstance.

We have already said that it is essential to have a comprehensive list of risks. It is very natural to introduce a typology of risks. There are numerous typologies, which make sense, and are helpful. We can compare risks of the same type, before comparing risks of different types. In industry one uses commonly the following typology:

- Strategic risks (survival of the company is at stake)
- Legal and reputational risks
- Financial risks
- Operational risks

It is only in the last type that one can find the technical risks. For banks we have commonly

- Financial markets
- Credit risk
- Operational risks

Although banks do not produce physical products, they have operations performed with people and information systems.

Proceeding with risk analysis requires identifying all risks, measuring them, and quantifying indicators. Indicators are built on structural randomness. This randomness is that which remains after all efforts to reduce the initial one have been implemented. Indeed, the initial randomness can be reduced by obtaining more information, or by any other measurement. Since the operation of obtaining more information is costly, at some point we decide that it is not worth the cost. The randomness which remains is structural randomness.

Note that opportunities are analyzed in the same way as risks.

Among the decisions there is that of accepting or not a risk. We face a psychological difficulty here. This risk can come from a project proposed by a team. The team may be frustrated by a negative decision. But accepting too many risks is also dangerous. We are in a situation called moral hazard. Insurance companies know this problem very well. People who insure their house may make less efforts to protect it. Drivers may become more adventurous if they feel safe by wearing a safety belt.

Performing a risk analysis and a risk management study remains a difficult task. There are objective reasons: we need to obtain data. Measuring risks requires expertise which may be rare. The information system may be complex. There are also objective reasons. There is resistance to change, lack of support of the hierarchy, power struggles, and discomfort with uncertainties.

If we perform such a study, the results must be correct. Wrong figures are worse than no figures. Finally, not everything is quantifiable. This is the case for political risks.

To conclude we emphasize the three pillars of Risk Management

- Quantitative techniques
- Psychological aspects
- Organizational aspects.

Omitting one pillar will have catastrophic consequences, whatever the attention given to others.

## 3.10 Case Studies

To illustrate all the above ideas, and particularly the concept of three pillars, we will discuss two cases, which, interestingly enough, relate to highly technological industrial sectors, Nuclear Industry and Aeronautics, to world leaders of the domain, in two of the most advanced countries in the world. We want to emphasize the commonalities of these two cases from the point of view of Risk Management, so that a scientific approach becomes meaningful and useful.

### *3.10.1 The Fukushima Accident*

On March 11, 2011, after an earthquake and a tsunami the tragedy of Fukushima broke out. Everyone keeps in mind that it is a nuclear accident, almost forgetting the earthquake and the tsunami, which are so common in Japan. The main message which spread out was that nuclear technology was very dangerous. Many countries, starting with Germany, decided to drop nuclear energy forever, or at least to reduce its use. Such a harsh decision, without a thorough analysis of the consequences, shows the lack of a basic understanding of Risk Management at the level of governments. But it is even more true at the level of the company TEPCO, in charge of the reactors. TEPCO is number one in Japan and one of the biggest companies in the world, in the nuclear industry.

Here was a prosperous company, well managed, at least with respect to the classical rules of management, one of the world leaders, which became in quasi bankruptcy within a few days. It could not have survived, without the intervention of the Japanese government. What is the use of maximizing its profit if everything is lost so quickly?

The obvious comment is that the company had no idea of what Risk Management is. Indeed, the accident was not a nuclear accident, but the consequence of failure in basic risk management. It is apparent, from looking at the chart of the company at the time of the accident, that no department or functional unit oversaw risk analysis or management. The closest was related to quality of the plants and of the energy produced. It means that TEPCO was concerned about the risks of its production process and about the quality of its products, in line with the concept of total quality, which characterizes Japanese industry, but nothing else.

As we have seen, the first and essential component of Risk Management is to identify all risks, certainly not just those related to the technology and the production process.

If one looks at figures, the misbehavior is obvious. The magnitude of the earthquake was 9, on the Richter scale. This is a very high figure, but the plants were designed to withstand a magnitude of 8.2 and thanks to the security margins did not collapse. So, the earthquake is not directly responsible. What is responsible is the tsunami which came afterwards. The tsunami wave was 14 m, which is high but not exceptional, and the protection wall was only 5.7 m high. Moreover, the plants are on the seashore and touching each other. Clearly, the nuclear plants had no protection against tsunamis. More troubling is the fact that many warnings about tsunami risks have been expressed, including from engineering staff inside the company itself.

To complete the landscape, TEPCO has another installation, called FUKUSHIMA II, further from the sea, which resisted the wave. Similarly, a competitor of TEPCO, which is smaller, Tohoku Electric Power Company is operating at Onagawa, closer to the epicenter of the earthquake, and had no accident, simply because the protection wall was 14.8 m high and resisted.

If the reasons for the accident have no ambiguity, one may wonder why the CEO and the Board of Directors were so unconscious of the tsunami risk. This looks like the blindness effect, described above. We also said that the CEO must be directly involved, otherwise there is no way that subordinates will compensate for the lack of leadership.

We have also in this case the failure of the regulatory agency, which did not correctly play its role. It also focused on the state of the plants and was too lenient with the industry, showing a lack of independence, which is too common, unfortunately.

### 3.10.2 The Boeing Case

A second interesting case is the story of the BOEING 737 Max 8. It starts with the crashes of LION Air 610, on 10-29-2018, and ETHIOPIAN Airlines 302, on 03-10-2019.

They were not unpredictable accidents. They should not have occurred. We deplore the death of 346 innocent people.

Why are those accidents very informative from a Risk Management point of view?

This aircraft was new and very important for Boeing. It was the flagship plane, introduced in March 2017, and supposed to represent one-third of the revenue of the company for at least 5 years.

Boeing is locked with Airbus in a race for the massively profitable market of middle-range planes. The issue is to offer airlines the cheapest airplane to operate. What is at stake is to reduce fuel consumption. Both companies have the same engine supplier, a consortium of GE and SAFRAN. The consortium maintains two lines of production, with no connection between them. We can assume that each company has some information on what the other one is preparing, but the decision to launch a new plane is a real secret. On December 1, 2010, Airbus stunned the aviation community by announcing the A320neo, which will burn 6% less fuel than the existing Boeing competitor, the 737 NG (Predecessor of 737 Max).

This is the origin of all the risky decisions of Boeing. Its hubris could not stand that Airbus takes the leadership in such an important market. Boeing's execs made up their minds in a matter of weeks. The company would launch a fourth-generation 737, and it would do it in record time. Using the 737 platform would save billions of dollars in engineering costs. With this choice, the narrative would be that, although as efficient as the competition, it was not a new plane. "Not a new plane" was very important, because the certification could be fast, and only a light training would be necessary for the pilots. It was an obsession, but also a myth.

They communicated a lot, without being sure to deliver. They announced that the Max will be 8% more efficient than the A320neo. Boeing is very trustworthy. They could sell a lot of planes and the FAA granted the max an amended type certification.

Unfortunately, they overlooked a very serious system engineering issue. The new engine, called LEAP-1B, is much more efficient than its predecessor, but at the same time much heavier and larger. It could not be mounted on the same spot as the previous one, because there would be too little clearance from the ground during take-off. So, the new engines were placed further forward and slightly higher on the wing of the max.

That solution created an aerodynamics problem. Due to their size and position, the engines on the max create lift when the airplane enters a steep climb (at high angles of attack). This extra lift causes the Max to behave differently than previous versions of the 737, supposedly only when it is climbing steeply.

The reasonable solution would have been to modify the platform. But then, the new plane would look significantly different from its predecessor, denying the narrative that the changes were limited.

So, they decided to solve a hardware difficulty by software. On paper, it was simple. A censor would detect when the airplane entered a steep climb, then the software would activate the airplane's pitch trim system to stabilize it until it detected that the steep climb was ended.

BOEING is not a software company. They use suppliers who do not have the responsibility to check how the software embedded in the plane behaves.

The software was activated by a single censor. It always believes that the information is correct.

Considering that the software will be needed rarely, they did not even mention it in the pilot's manual. Even more serious, when activated the software overrides the actions of the pilot.

Risks for embedded software are technical risks, which are known and require adapted methods. None of them were implemented. Yet warnings have been signaled by test pilots using a flight simulator.

As mentioned, there were two crashes, with more than 6 months between the two.

Instead of acknowledging the issue after the first crash, BOEING minimized it, even blaming the pilots, and promising an easy fix. In addition, like in the Fukushima accident, the regulatory agency, the FAA, failed in exerting its responsibility.

## 3.11 Conclusion

The study of uncertainties and of decisions can be considered as a science, since there is a substantial number of concepts and methods with a broad range of applications. Risk Management is its engineering counterpart. New risks have introduced new challenges, because of their psychological aspects, which cannot just be addressed with technology and mathematics. Organization is needed.

## References[1]

Bernstein, P. L. (1998). *Against the gods: The remarkable story of risk*. Wiley.
Vose, D. (2008). *Risk analysis: A quantitative guide* (3rd ed.). Wiley.
The first one presents the state of the art, the second is historical.
The author has published in French: Science des risques et décision, Risques, n. 107, pp. 135–137, 2016.

[1] Literature is now very abundant. We recommend below:

# Chapter 4
# Optimization in Age-Structured Dynamic Economic Models

**Michael Freiberger, Michael Kuhn, Alexia Prskawetz, Miguel Sanchez-Romero, and Stefan Wrzaczek**

**Abstract** Age-structured optimal control models experience increasing applications in various research fields including, e.g., demography, economics, operations research, epidemiology, and environmental economics. In this paper we present the mathematical theory and potential applications of age-structured optimal control models. We first state the general form of the problem and present the necessary optimality conditions. To illustrate the mathematical theory we introduce a toy model on air pollution, where consumption induces pollution which in turn negatively affects utility, fertility, and mortality. We solve the model analytically and present numerical simulations. The potential of an age-structure approach to solve non-standard optimal control models is demonstrated by considering optimal control models with random switches or time-lags and delays.

---

M. Freiberger
International Institute for Applied Systems Analysis (IIASA), Laxenburg, Austria

Wittgenstein Center (IIASA, OeAW/VID, University of Vienna), Wien, Austria

M. Kuhn
International Institute for Applied Systems Analysis (IIASA), Laxenburg, Austria

Wittgenstein Center (IIASA, OeAW/VID, University of Vienna), Wien, Austria

Vienna Institute of Demography, Wien, Austria

A. Prskawetz · M. Sanchez-Romero
Research Group Economics, Institute of Statistics and Mathematical Methods in Economics, Vienna University of Technology, Vienna, Austria

Vienna Institute of Demography (VID), Austrian Academy of Sciences, Wien, Austria

International Institute for Applied Systems Analysis (IIASA), Laxenburg, Austria

Wittgenstein Center (IIASA, OeAW/VID, University of Vienna), Wien, Austria

S. Wrzaczek (✉)
International Institute for Applied Systems Analysis (IIASA), Laxenburg, Austria

Wittgenstein Center (IIASA, OeAW/VID, University of Vienna), Wien, Austria
e-mail: wrzaczek@iiasa.ac.at



F. El Ouardighi, G. Feichtinger (eds.), *The Unaffordable Price of Static Decision-making Models*, International Series in Operations Research & Management Science 365, https://doi.org/10.1007/978-3-031-88638-6_4

**Keywords** Age-structured optimal control · McKendrick-von Foerster equation · Non-standard optimal control · Random switch · Optimal control with time-lag or delay

## 4.1 Introduction

In optimal control models (or dynamic optimization models, in general) the dynamics describe the evolution of a system along the direction of an *independent* variable, which typically is either time $t$, as, e.g., in macroeconomic planning problems, or age $a$, as in microeconomic life-cycle problems. The dynamics are modeled by a system of (first-order) ordinary differential equations (or difference equations). However, in applications that require in-depth models of the dynamics of a population—such as the modeling of social security, labor market, and health policies—as well as applications relating to epidemiology, harvesting, and the employment of capital vintages, age becomes a crucial variable in addition to and in distinction to time. The key equation that models the dynamics along time and age is a first-order partial differential equation, i.e.,

$$y_t(t,a) + y_a(t,a) = f(\cdot), \qquad y(0,a) = y^0(a), \quad y(t,0) = \varphi(\cdot), \qquad (4.1)$$

known as the McKendrick-von Foerster equation (see, e.g., Keyfitz and Keyfitz, 1997). Here, $y(t,a)$ denotes the state variable at time $t$ and age $a$. $y_t(t,a)$ and $y_a(t,a)$ denote the partial derivative of $y(t,a)$ with respect to time and age, respectively. Thus, the left-hand side of (4.1) denotes the directional derivative of $y(t,a)$ in direction $(1,1)$ since time and age evolve at the same pace. Details of the right-hand side of the equation, as well as of the initial and boundary conditions, will be discussed in the next section. In addition to (4.1) an age-structured optimal control model allows the objective and salvage value functions to be age-structured, as well as *aggregated* state variables $Q(t)$ to be included. These variables aggregate/integrate (system) effects across (all) age groups at any given point in time.

The literature on age-structured optimal control theory evolved as a sequence of papers deriving a maximum principle (MP) for a specific problem. The first rigorous proof for a general setup with a nonlinear McKendrick-von Foerster equation was presented by Brokate (1985). Feichtinger et al. (2003) generalized the maximum principle by adding an age-dependent aggregated state variable. Veliov (2008) provided a MP for general heterogeneous systems.

Applications of age-structured optimal control theory are broad and originally emerged in population dynamics and population economics; see Arthur and McNicoll (1977), Chan and Guo (1989, 1990), Gurtin and Murphy (1981a,b), Medhin (1992), Feichtinger et al. (2004), Prskawetz and Veliov (2007), Feichtinger et al. (2012), Prskawetz et al. (2012), or Feichtinger and Wrzaczek (2024a,b). In parallel the theory was also applied in the mathematical literature on epidemiology (see

Greenhalgh, 1988; Hethcote, 1988) and economics (see Derzko et al., 1980; Feichtinger et al., 2006; Kuhn et al., 2011; Augeraud-Véron et al., 2019; Hartl et al., 2023), among other fields.

The contribution of this paper to the optimal control literature is twofold. First, we formulate the age-structured MP in a general abstract way and show how it is used in a toy model on air pollution. Within this model we also elaborate how the age structure enters the necessary conditions (of the MP), how it changes the solution in comparison to a standard (time-dependent) optimal control model, and how it can be understood in an intuitive way. In so doing, we seek to create an understanding of the relevance of the age-time dynamic in optimal control theory for addressing important policy questions. Second, apart from the importance of age structure as a dynamic dimension, we demonstrate how the age-structured MP can be used to handle advanced non-standard optimal control models that, otherwise, are difficult to deal with. The transformation uses age structure as auxiliary dimension but substantially improves (intuitive) insights and facilitates the solution of model classes that to date are applied to a limited extent only.

The paper is organized as follows. Section 4.2 presents the age-structured MP, which is applied in Sect. 4.3 to a toy model on air pollution. Section 4.4 discusses how age structure can be used for non-standard (advanced) optimal control models. Section 4.5 concludes.

## 4.2 The Age-Structured Maximum Principle

Let us first state the general form of an age-structured optimal control problem. In the following problem (4.2a) denotes the objective function, (4.2b) and (4.2c) the model dynamics, and (4.2d) and (4.2e) the initial and boundary conditions:

$$\max_{\substack{u(t,a)\in U\\ v(t)\in V}} \int_0^T \int_0^\omega L\left(y(t,a), Q(t), u(t,a), v(t), t, a\right) \mathrm{d}a\mathrm{d}t$$

$$+ \int_0^\omega S\left(y(T,a), T, a\right) \mathrm{d}a \tag{4.2a}$$

$$\text{s.t.}\;\; y_t(t,a) + y_a(t,a) = f(y(t,a), Q(t), u(t,a), v(t), t, a) \tag{4.2b}$$

$$Q(t) = \int_0^\omega h\left(y(t,a), Q(t), u(t,a), v(t), t, a\right) \mathrm{d}a \tag{4.2c}$$

$$y(0,a) = y^0(a) \tag{4.2d}$$

$$y(t,0) = y^b(Q(t), v(t), t). \tag{4.2e}$$

Here, $t$ and $a$ denote time and age, respectively, with time horizon $T$ and maximal attainable age $\omega$. $y(t,a) \in \mathbb{R}^m$ and $Q(t) \in \mathbb{R}^n$ are distributed and aggregated state variables.[1,2] The corresponding functions $f$ and $h$ depend on time, age, state variables, and the control variables denoted by $u(t,a) \in U \subseteq \mathbb{R}^p$ (distributed) and $v(t) \in V \subseteq \mathbb{R}^q$ (concentrated). While $y^0(a)$ denotes the exogenous[3] initial distribution of $y(t,a)$ across age at time $t = 0$, $y^b(Q(t), v(t), t)$ denotes the boundary condition which can depend on the aggregated state as well as on the concentrated control variables. In contrast to $y(t,a)$, an initial or boundary condition is not required for $Q(t)$, as it is derived from the aggregation of $h(\cdot)$ at every $t$. The decision maker chooses $u(t,a)$ and $v(t)$ in order to maximize the sum of the aggregated objective $L(\cdot)$ and salvage value function $S(\cdot)$ (see (4.2a)).

The Lexis diagram shown in Fig. 4.1 illustrates how variables in the general model (4.2) relate to the time and age dimension in the model. Time and age are plotted on the horizontal and vertical axes, respectively. The characteristic lines (45° lines in blue) show that time and age evolve at the same pace; hence, $t - a$ denotes the time at which a specific characteristic line emerges. $y(t,a)$ and $u(t,a)$ are time- and age-specific variables. They evolve along characteristic lines and emerge either at the vertical axes according to the initial condition $y^0(a)$ or at the horizontal axes according to the boundary condition $y^b(Q(t), v(t), t)$. $Q(t)$ results from an aggregation across the age domain at $t$ and influences the dynamics (4.2b)–(4.2c), the boundary condition (4.2e), and the objective function (4.2a). Thus, the Lexis diagram highlights the asynchrony of the variables with respect to the time and age domain, which is the intuitive reason for a separate MP for these problems.

The length of the time horizon $T$ and the maximal attainable age $\omega$ are both finite and define the intervals $D^T := [0, T]$, $D^A := [0, \omega]$, as well as the domain $D := D^T \times D^A$ within which the distributed state and control variables are defined. Note that this is in line with most theoretical and applied works based on age-structured optimal control models. In contrast to time-dependent optimal control models, the formulation of general limiting transversality conditions for the adjoint variables is difficult for age-structured optimal control models and, therefore, implies the absence of a general MP for problem (4.2) with infinite time horizon.[4]

[1] Note that the MP presented in Feichtinger et al. (2003) also allows for distributed aggregate state variables. For an application see, e.g., Almeder et al. (2004).

[2] Note that problem (4.2) can easily be extended to include a concentrated state variable $x(t)$, whose dynamic is described by an ordinary differential equation (ODE). This is important in a number of applications such as the employment of age structure in multi-stage optimal control models with stochastic switches and optimal control models with time-lag as discussed in Sect. 4.4. For the extended necessary conditions and a sketch of a proof, we refer to Feichtinger and Wrzaczek (2024a).

[3] Note that the MP of Feichtinger et al. (2003) also allows the control of the initial condition $y^0(a)$ by a purely age-dependent control variable. This is similar to a control of the initial condition in a standard optimal control model. Due to its infrequent usage, a control variable of this type is not presented here.

[4] For a discussion on a MP for infinite time horizon for the specific case of the PDE being linear in $y(t,a)$, see, e.g., Skritek and Veliov (2015) or Buratto et al. (2020).

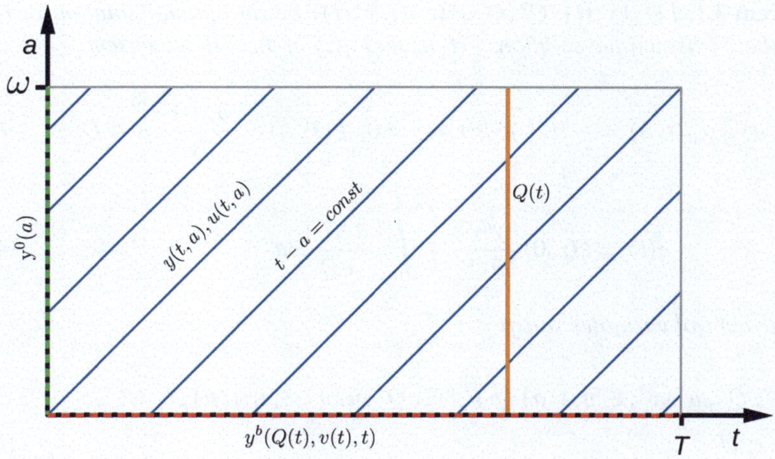


**Fig. 4.1** Lexis diagram: showing variables and conditions along time and age dimension

**Table 4.1** Variables, functions, and conditions of an age-structured optimal control model (4.2)

| | | |
|---|---|---|
| Independent variables | Time | $t \in D^T$ |
| | Age | $a \in D^A$ |
| Control variables | Distributed | $u(t,a) : D \mapsto U$ |
| | Concentrated | $v(t) : D^T \mapsto V$ |
| State variables | Distributed | $y(t,a) : D \mapsto \mathbb{R}^m$ |
| | Aggregated | $Q(t) : D^T \mapsto \mathbb{R}^n$ |
| Functions | Objective functional | $L : \mathbb{R}^m \times \mathbb{R}^n \times U \times V \times D \mapsto \mathbb{R}$ |
| | Salvage value | $S : \mathbb{R}^m \times D \mapsto \mathbb{R}$ |
| | Distributed system dynamic | $f : \mathbb{R}^m \times \mathbb{R}^n \times U \times V \times D \mapsto \mathbb{R}^m$ |
| | Aggregation | $h : \mathbb{R}^m \times \mathbb{R}^n \times U \times V \times D \mapsto \mathbb{R}^n$ |
| Initial and boundary conditions | Initial distribution of $y(0,a)$ | $y^0 : D^A \mapsto \mathbb{R}^m$ |
| | Boundary condition of $y(t,0)$ | $y^b : \mathbb{R}^n \times V \times D^T \mapsto \mathbb{R}^m$ |

Table 4.1 summarizes the variables and functions that define (4.2). We assume the admissible control sets to be compact and convex, and the involved functions to be twice continuously differentiable.[5]

The age-structured MP formulates necessary optimality conditions for problem (4.2).

[5] Note that these assumptions are stronger than necessary. We refer to Brokate (1985) or Feichtinger et al. (2003) for weaker assumptions that are sufficient to prove the MP.

**Theorem 4.1** *Let $(y^*(t), Q^*(t), u^*(t,a), v^*(t))$ be an optimal solution of (4.2). Then there exist unique solutions $\xi(t,a)$ and $\eta(t)$ of the adjoint system*

$$\xi_t(t,a) + \xi_a(t,a) = -\mathcal{H}_y(\cdot),\ \xi(t,\omega) = 0,\ \xi(T,a) = \frac{\partial S(\cdot)}{\partial y},\ a \in D^A, t \in D^T \tag{4.3a}$$

$$\eta(t) = \xi(t,0)\frac{\partial y^b(\cdot)}{\partial Q} + \int_0^\omega \frac{\partial \mathcal{H}(\cdot)}{\partial Q} da \tag{4.3b}$$

*and the control variables satisfy*

$$\mathcal{H}\left(y^*, Q^*, u^*, v^*, \xi, \eta, t, a\right) \geq \mathcal{H}\left(y^*, Q^*, u, v^*, \xi, \eta, t, a\right),\quad \forall u \in U \tag{4.4a}$$

$$\left(\xi(t,0)\frac{\partial y^b}{\partial v}(Q^*, v^*, t) + \int_0^\omega \frac{\partial \mathcal{H}}{\partial v}\left(y^*, Q^*, u^*, v^*, \xi, \eta, t, a\right) da\right)\left(v^* - v\right) \geq 0,\quad \forall v \in V \tag{4.4b}$$

*for a.e. $t \in D^T$ and $(t,a) \in D$, where the Hamiltonian is defined by*

$$\begin{aligned}\mathcal{H}\left(y^*, Q^*, u^*, v^*, \xi, \eta, t, a\right) := {} & L\left(y(t,a), Q(t), u(t,a), v(t), t, a\right) \\ & + \xi(t,a) f\left(y(t,a), Q(t), u(t,a), v(t), t, a\right) + \\ & + \eta(t) h\left(y(t,a), Q(t), u(t,a), v(t)\right).\end{aligned} \tag{4.5}$$

Here $\xi(t,a)$ and $\eta(t)$ denote the adjoint variables corresponding to the state variables $y(t,a)$ and $Q(t)$, respectively. They share the dimension and the same dependencies with their corresponding state variables.[6] A strict proof of Theorem 4.1 can be found in Brokate (1985), Feichtinger et al. (2003) (including a distributed aggregated state variable), or Veliov (2008) (for the case of more general heterogeneous systems). Feichtinger and Wrzaczek (2024a) explicitly formulate the optimality conditions with an additional concentrated state variable (which is a special case of the previous papers) and provide a sketch of the proof. Wang (1964) and Brogan (1968) derive the MP by using a dynamic programming approach, which allows interpreting the adjoint variables as a shadow price.

The MP in Theorem 4.1 is formulated in a general way. For an interior solution of the control variables Equation (4.4) reduces to

[6] Note that the multidimensional adjoint variables should be read as row vectors, while the state and control variables are column vectors by definition.

$$\frac{\partial \mathcal{H}(\cdot)}{\partial u} = 0 \tag{4.6a}$$

$$\xi(t,0)\frac{\partial y^b(\cdot)}{\partial v} + \int_0^\omega \frac{\partial \mathcal{H}(\cdot)}{\partial v} da = 0. \tag{4.6b}$$

To obtain the maximum of the Hamiltonian (where (4.2) is formulated as a maximization problem), the (static) second-order condition, not explicitly formulated here, has to be fulfilled as well. Note that (4.6a) has to hold for every $(t, a)$, whereas (4.6b) corresponds to $t$ alone, implying that the derivative of the Hamiltonian with respect to $v(t)$ is aggregated across the age domain.

Analogous to the MP for time-dependent optimal control models, the age-structured MP provides a set of conditions that are necessary for optimality. General sufficiency conditions akin to the Arrow, the Mangasarian, or the Leitmann-Stalford conditions (cf. Seierstad and Sydsaeter, 1977 and Leitmann and Stalford, 1971) are not available so far and have to be derived in relation to the specific problem.

Compared to time-dependent optimal control models, the numerical treatment of age-structured optimal control problems is substantially more involved. As presented above, the necessary optimality conditions consist of a set of partial differential equations (PDEs) combined with boundary conditions for state and co-state variables as well as algebraic equations for all $(t, a)$. In general, the solution process for a set of PDEs is highly complex already. However, $t - a =$const for the PDEs in an age-structured optimal control model allows the use of the methods of characteristics (see Zachmanoglou and Thoe, 1986). This solution technique reduces each PDE to a set of ordinary differential equations (ODEs). Each ODE represents a cohort and it can be solved numerically with a wide range of established solution techniques, significantly reducing the degree of difficulty/complexity of the numerical problem.

Nevertheless, the difficulties resulting from the mixed boundary conditions (initial conditions for the state variables/end conditions for the co-state variables) and the algebraic optimality conditions remain. We will now briefly discuss two potential iterative approaches to solve these issues: (i) shooting algorithms and (ii) gradient-based algorithms.

Shooting algorithms[7] start with a guess for the initial values of the co-state variables. The state and co-state variables are then solved forward in time with the algebraic equations being solved (analytically or numerically) at each point in time to obtain values for the control variables. According to the discrepancy between the end values of the co-state variables and the target values according to the boundary conditions, the initial values for the co-state variables are adjusted iteratively until the end conditions for the co-state variables are fulfilled (within a given margin of error).

[7] See Bonnans (2013) for an overview of shooting algorithms for optimal control problem.

Conversely, gradient-based algorithms[8] start with an initial guess for the control variables over the whole domain. Using this guess the state variables are calculated iterating forward in time. Given the state profiles the end constraints for the state variables can be evaluated and co-state dynamics are solved backward in time (starting at $t = T$ and ending at $t = 0$). Given this solution for states, co-states, and controls, the gradient of the Hamiltonian is evaluated and the controls are adjusted in the direction of the gradient to find an improvement in the objective function. These steps are iterated until no further improvement in the objective function is found.

Each approach has its own advantages and disadvantages with respect to computation times or stability and range of convergence, but both can provide a solution of the full system.

## 4.3 Toy Model on Air Pollution

We take inspiration from recent work on the pathways and welfare impacts of consumption-based air pollution (e.g., Zhao et al., 2019; Almetwally et al., 2020; Rao et al., 2021; Peszko et al., 2023) for the purpose of illustrating the advantages of considering an age-structured population within an optimal control model. Specifically, we employ the model to derive the welfare-maximizing allocation of consumption across a population and over time when taking into account negative impacts of consumption-driven pollution on health and productivity.

The economy consists of an age-structured population $N(t, a)$, the dynamics of which are driven by an age-specific mortality rate $\mu(\cdot)$ and fertility rate $\nu(\cdot)$, both of which depend on pollution $P(t)$. In line with the above-cited literature, pollution is assumed to increase mortality for all age groups; fertility is negatively affected by pollution (e.g., Conforti et al., 2018 and Jurewicz et al., 2018 on biomedical channels (fecundity) and Gao et al., 2022 on socioeconomic channels). The initial population structure at time $t = 0$ is exogenously given by $N_0(a)$; the number of births $B(t)$ defines the population of age $a = 0$ at every $t$ and results from the total fertility of the population. These population dynamics are summarized in Eqs. (4.7b), (4.7c), and (4.7f).

The cohort of age $a$ at time $t$ holds a total value of $A(t, a)$ in assets. These assets generate interest at the rate $r$ and are adjusted at every point in time $t$ by the difference between age-specific (per capita) earnings $w(a, P(t))$, also assumed to depend negatively on pollution (e.g., Aguilar-Gomez et al., 2022; Neidell, 2023), and consumption $c(t, a)$. Individuals start their lives with zero assets ($A(t, 0) =$

[8] See Veliov (2003) for the theoretical proof of convergence of the Newton's method for age-structured optimal control problems.

0) and have to possess zero assets at their maximum age of survival $\omega$.[9] For the dynamic and boundary equations for cohort assets, see Eqs. (4.7d) and (4.7e).

Air pollution is assumed to be a flow variable in our toy model and results from the total consumption across all cohorts (see (4.7g)).

The planner's objective is to maximize social welfare, which is defined by the total utility aggregated across time and cohorts. The per capita period utility function $u(c(t,a), P(t))$ increases with per capita consumption $c(t,a)$ and decreases with the total pollution in the economy, the latter reflecting direct negative effects on physical or mental health (e.g., Almetwally et al., 2020; Shi and Yu, 2020). The objective function in (4.7a) is of the Benthamite type and counts the utility of every individual at $t$.[10] The model can be summarized as follows:

$$\max_{c(t,a)} \int_0^T \int_0^\omega e^{-\rho t} N(t,a) u(c(t,a), P(t)) \mathrm{d}a \mathrm{d}t \tag{4.7a}$$

$$\text{s.t. } N_a(t,a) + N_t(t,a) = -\mu(a, P(t)) N(t,a), \tag{4.7b}$$

$$N(0,a) = N_0(a), \quad N(t,0) = B(t) \tag{4.7c}$$

$$A_a(t,a) + A_t(t,a) = rA(t,a) + (w(a, P(t)) - c(t,a)) N(t,a), \tag{4.7d}$$

$$A(0,a) = A_0(a), A(t,0) = 0, A(T,a) = A(t,\omega) = 0 \tag{4.7e}$$

$$B(t) = \int_0^\omega \nu(a, P(t)) N(t,a) \mathrm{d}a \tag{4.7f}$$

$$P(t) = \int_0^\omega c(t,a) N(t,a) \mathrm{d}a. \tag{4.7g}$$

[9] Note that defining $A(t,a)$ as cohort assets (rather than per capita assets) allows us to easily incorporate the redistribution of assets that are held by deceased individuals. In our toy model the assets automatically get redistributed between the surviving individuals of the same cohort. This fact becomes obvious when examining the differential equation for per capita assets $S(t,a) := A(t,a)/N(t,a)$.

$$S_t(t,a) + S_a(t,a) = (r + \mu(a, P(t))) \cdot S(t,a) + w(a, P(t)) - c(t,a).$$

This equation shows that assets get redistributed equivalently to an annuity market that covers the mortality risk.

[10] While in the Benthamite setting, per capita utility is scaled with the cohort size $N(t,a)$, the Millian social welfare function is based on the average utility across the whole population (see, e.g., Kuhn et al., 2010) and can straightforwardly be obtained by dividing utility by the total population size at time $t$.

### 4.3.1 Analysis and Economic Insight

We now demonstrate the age-structured MP by following Theorem 4.1. The current value Hamiltonian (Eq. (4.5)) reads (ignoring $t$ and $a$ for simplicity)

$$\mathcal{H} = Nu(c, P) + \xi^N \left(-\mu(P)N\right) + \xi^A \left(rA + (w(P) - c)N\right) + \eta^B v(P)N + \eta^P cN, \tag{4.8}$$

where $\xi^N(t, a)$ and $\xi^A(t, a)$ denote the adjoint variables of the (distributed) population and asset states, and where $\eta^B(t)$ and $\eta^P(t)$ denote the adjoint variables of the (aggregated) births and pollution states, respectively. As an implication the necessary first-order conditions for age-structured consumption follow from Eqs. (4.4a) and (4.6a), i.e.,

$$\begin{aligned} \mathcal{H}_c &= Nu_c - \xi^A N + \eta^P N = 0 \\ &\Longrightarrow u_c = \xi^A - \eta^P, \quad (t, a) \in D. \end{aligned} \tag{4.9}$$

Equations (4.4b) and (4.6b) are not used, since the toy model does not include a concentrated control variable. Equation (4.9) reflects the standard *marginal utility* = *marginal costs* criterion in economics. The left-hand side (lhs) equals the marginal utility of an individual (aged $a$ at $t$). The right-hand side (rhs) comprises the value of assets, reflecting alternative future consumption possibilities, and the shadow cost of air pollution, embracing an immediate effect on instantaneous utility and an intertemporal effect on mortality and fertility. Note here that typically $\xi^A$ is positive and $\eta^P$ is negative. Here, the rhs also illustrates the interaction of the independent dimensions *age* and *time*. Whereas $\xi^A$ (depending on age and time) depicts the intertemporal effect along the life-course of a cohort, $\eta^P$ evaluates the cost of pollution across all cohorts at $t$ and intertemporally. The social optimum, thus, includes cross-cohort pollution damages to the optimization nexus of the age-structured consumption decision. This feature cannot be obtained in a standard (i.e., concentrated parameter) optimal control model that is simplified by neglecting the age dimension.

The adjoint equations and transversality conditions are derived straightforwardly (Eqs. (4.3)). We obtain

$$\xi^N_t + \xi^N_a = (\rho + \mu)\,\xi^N - u - \xi^A (w - c) - \eta^B v - \eta^P c \tag{4.10a}$$

$$\xi^A_t + \xi^A_a = (\rho - r)\,\xi^A \tag{4.10b}$$

$$\eta^B = \xi^N(t, 0) \tag{4.10c}$$

$$\eta^P = \int_0^\omega \left(Nu_P - \xi^N \mu_P N + \xi^A w_P N + \eta^B v_P N\right) da, \tag{4.10d}$$

with

$$\xi^N(T,a) = 0, \qquad a \in D^A \tag{4.11a}$$

$$\xi^N(t,\omega) = 0, \qquad t \in D^T. \tag{4.11b}$$

Regarding the adjoint variables for the aggregated state variables (4.10c)–(4.10d), we would like to emphasize the structural difference between $\eta^B$ on the one hand and $\eta^P$ on the other hand, although all of these are derived from the general expression (4.3b). Births $B(t)$ do not enter the objective function and the system dynamics, but only the boundary condition of the population. This means that $B(t)$ is affecting $N(t,a)$ only once (i.e., at $a = 0$) and is covered by the first term of (4.3b). Pollution $P(t)$, on the other hand, enters the objective function and/or the system dynamics but not the boundary condition of the population. Therefore, (4.10d) draws on the second term of (4.3b) covering the effect across all cohorts at $t$ and intertemporally.

To explore the dynamics of an optimal allocation, the derivative of the control variable (starting from the first-order condition), in economics referred to as *Euler equation*, can be used. In the case of an age-structured control variable, the derivative has to be taken along time and age. Using (4.9) and (4.10) we obtain the following general expression:

$$\frac{c_t + c_a}{c} = -\frac{u_c}{u_{cc} \cdot c}\left((r-\rho) + \frac{(r-\rho)\,\eta^P + \eta_t^P}{u_c} + P_t\frac{u_{cP}}{u_c}\right). \tag{4.12}$$

The equation determines whether it is better to postpone or advance consumption. The first term on the rhs outside the parenthesis shows the social planner's absolute risk aversion or, equivalently, the inverse of the elasticity of intertemporal substitution. A more risk-averse social planner is less responsive to changes in the economy and less willing to shift consumption over time. The first term inside the parenthesis shows the difference between the current valuation of savings by the market ($r$) and the social planner ($\rho$). If the market values savings more (or less) than the social planner, i.e., $r > \rho$ (or $r < \rho$), the social planner has an incentive to defer (or advance) consumption, implying an increase (decline) in consumption over time. The second term inside the parenthesis depicts how the social planner values the evolution of pollution. Noting that pollution typically carries a negative value, i.e., that $\eta^P < 0$, a further decrease (increase) toward a more (less) negative value, i.e., $(r-\rho)\eta^P + \eta_t^P < (>)0$, implies that the social planner chooses to advance (postpone) consumption and reduce (or increase) savings. This is to reduce future (present) pollution damage. The third term inside the parenthesis accounts for the negative impact on the utility from consumption of increasing pollution. Assuming a negative impact $u_{cP} < 0$, an increase (decrease) in the pollution flow over time implies an advancement (deferral) of consumption.

As the shadow price of pollution $\eta_P$ itself is partially determined by the shadow price of the population, it is helpful to consider the analytic expression of $\xi^N(t,a)$

obtained by backward integration of (4.10a). Using (4.11b) we obtain the following expression for an individual that dies before $T$ (i.e., $t - a \leq T - \omega$):[11]

$$\xi^N(t,a) = \int_a^{\omega} e^{-\int_a^s (\rho+\mu)\mathrm{d}s'} \left( \underbrace{u}_{(i)} + \underbrace{\left(u_c + \eta^P\right)(w-c)}_{(ii)} + \underbrace{\xi^N(t,0)\nu}_{(iii)} + \underbrace{\eta^P c}_{(iv)} \right) \mathrm{d}s. \tag{4.13}$$

The integral (4.13) aggregates the marginal effects on social welfare over the remaining lifespan of an individual aged $a$ at time $t$ discounted by $\rho$ and the conditional survival probability.[12] Therefore, the closed-form (4.13) can also be referred to as the *expected* present value of a consumer aged $a$ at time $t$ akin to similar expressions derived in Kuhn et al. (2010, 2011).

The present value of an additional consumer aged $a$ can be decomposed into four substantive parts (each discounted and weighed by the respective survival function). (i) denotes the increase in utility associated with this consumer. (ii) depicts the marginal effect (positive for utility, negative for pollution) of redistributing income over the consumer's remaining lifetime. This "cohort redistribution" effect obviously cannot be obtained in a dynamic model without age structure and includes the consumption path (along the life-cycle) as well as age-structured mortality and fertility. Term (iii) is a population dynamic effect that again can only be derived for an age-structured population. It captures the value of the expected progeny (as consumers) born to an individual born at $t - a$ over its own remaining life-cycle (including the effect of a newborn cohort on their offspring). This is a generalization of the demographic reproductive value and can be proven to appear in all age-structured optimal control models that model population via the McKendrick-von Foerster equation with endogenous births (see Kuhn et al., 2010; Wrzaczek et al., 2010; and Feichtinger et al., 2011). (iv) assigns to the individual the (negative) value of the pollution it causes over its remaining life-course. In this way the decision maker is able to internalize the cross-cohort pollution externality within and across cohorts (and over time).

### 4.3.2 Numerical Solution

For the numerical solution of the toy model, we use specific functional forms for the utility, the mortality rate, and the wage rate. The fertility rate is assumed not

[11] For an individual that is alive at $T$ (i.e., $t - a > T - \omega$) the expression is analogous, only with the upper bound of the interval now defined by the individual age at $T$ instead of $\omega$ and the employment of (4.11a) instead of (4.11b).

[12] Note that $e^{-\int_a^s (\rho+\mu)\mathrm{d}s'}$ can also be written in terms of state variables $\frac{N(t-a+s,s)}{N(t,a)}$.

to depend on pollution and to be equal to a standard baseline rate. In the following we briefly discuss the functional choices. The specific parameters can be found in Table 4.2.

For the utility we use a standard constant relative risk aversion (CRRA) function, which is multiplicatively reduced according to $e^{-\kappa_1(a)P^\gamma}$. The non-negativity of the exponential function guarantees a non-negative utility. The mortality rate is the baseline mortality rate, which follows a Gompertz law with a modal age at death of 80 years and a senescence rate of 0.10 (see Horiuchi et al., 2014), augmented by the effect of pollution. For the illustration purpose of the toy model, a linear form is sufficient. The wage rate is structured analogously, i.e., a baseline rate is reduced by a linear effect of pollution.

The initial population distribution is determined by stable population theory (see Coale, 1957), using the population growth rate as the ratio of the logarithm of the net reproduction rate to the average childbearing age. The wage rate is modeled by a standard Mincerian equation.

Although an age pattern of the pollution effects would be realistic and one reason for age dependence of the consumption profile, we assume age independence, i.e., $\frac{\partial \kappa_i(a)}{\partial a} = 0,\ i = 1, 2, 3$. Already in this simplified setup a nontrivial dynamic consumption path is optimal due to the interaction of the time and age domains within the problem. This observation would potentially be overlaid by an age dependence.

The model is calculated for 250 years with a maximal lifetime of 100 years. The remaining model parameters and the complete set of specific functional forms are listed in Table 4.2.

**Table 4.2** Summary of functions and parameters for numerical solution

| Function | Form | Parameters | Value |
|---|---|---|---|
| Base parameters | Discount rate | $\rho$ | 0.02 |
| | Market interest rate | $r$ | 0.02 |
| | Time horizon | $T$ | 250 |
| | Maximal age | $\omega$ | 100 |
| Utility | $u(c, P) = \left(b + \frac{c^{1-\sigma}}{1-\sigma}\right) e^{-\kappa_1(a)P^\gamma}$ | $b$ | 3 |
| | | $\sigma$ | 1.0 |
| | | $\gamma$ | 1.1 |
| | | $\kappa_1(a)$ | $1.5 \cdot 10^{-5}$ |
| Mortality rate | $\mu(a, P) = \tilde{\mu}(a)(1 + \kappa_2(a)P)$ | $\kappa_2(a)$ | $1.5 \cdot 10^{-5}$ |
| | | $\tilde{\mu}(a)$ | Calibrated |
| Fertility rate | $\nu(a) = \tilde{\nu}(a)$ | $\tilde{\nu}(a)$ | Calibrated |
| Wage rate | $w(t, a) = \tilde{w}(t, a) \cdot (1 - \kappa_3(a)P)$ | $\kappa_3(a)$ | $2.0 \cdot 10^{-4}$ |
| | | $g$ | $1.5 \cdot 10^{-3}$ |
| | | $\beta_0$ | $-6.66 \cdot 10^{-1}$ |
| | $\tilde{w}(t, a) = e^{\beta_0 + gt + \beta_1 a + \beta_2 a^2}$ | $\beta_1$ | $6.02 \cdot 10^{-2}$ |
| | | $\beta_2$ | $-9.0 \cdot 10^{-4}$ |

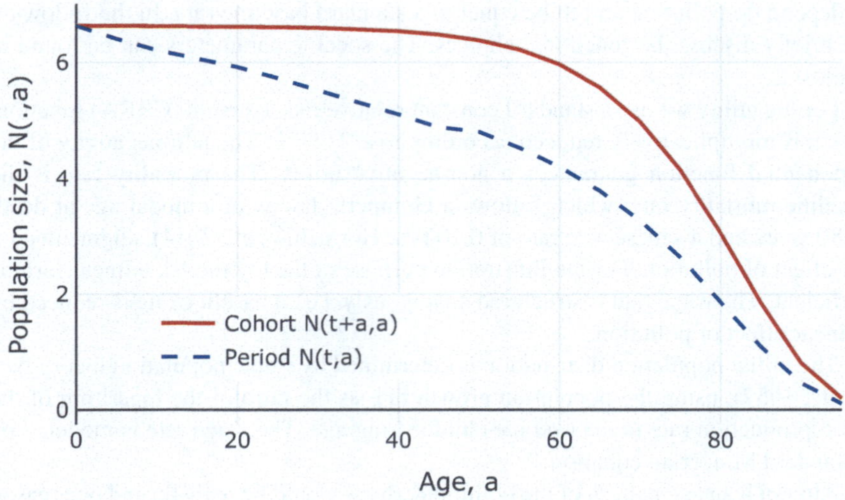


**Fig. 4.2** Population along a cohort (red line) and across cohorts (blue dashed line) for a fixed $t$

Figure 4.2 shows the age distribution of the population $N(t, a)$. The red line refers to the density of the cohort born at $t = 50$. The shape naturally follows the survival profile of an individual, which in our model corresponds to the base mortality rate augmented by negative pollution effects. The blue dashed line shows the population density across ages at the time of birth of the cohort shown with the red line, i.e., $t = 50$. As we assume a stable increasing population, the dashed blue line lies below the red one. The figure illustrates the possibility of modeling state (and control) variables in the two independent directions time and age, along which they develop differently. Whereas the red line corresponds to the 45° line in the Lexis diagram (Fig. 4.1), the blue dashed line resembles a vertical line in the Lexis diagram. Both of them start at the same time and age specific value $(t, a) = (50, 0)$.

Figure 4.3 plots the optimal consumption profile along the same dimensions as in Fig. 4.2, i.e., along the lifetime of one cohort (red line) and across cohorts for fixed $t$ (blue dashed line).

To discuss the shape of the red line, we refer back to the Euler equation (4.12). As the market interest rate and the social planner's time discount rate coincide ($r = \rho$), Eq. (4.12) simplifies to

$$\frac{c_t + c_a}{c} = \frac{1}{\sigma} \cdot \left( \frac{\eta_t^P}{u_c} - \kappa_1(a) \cdot \gamma \cdot P^{\gamma-1} \cdot P_t \right). \tag{4.14}$$

Equation (4.14) shows that the consumption profile would be flat, if pollution had none of the three externality effects, i.e., if it did not affect the utility, wage,

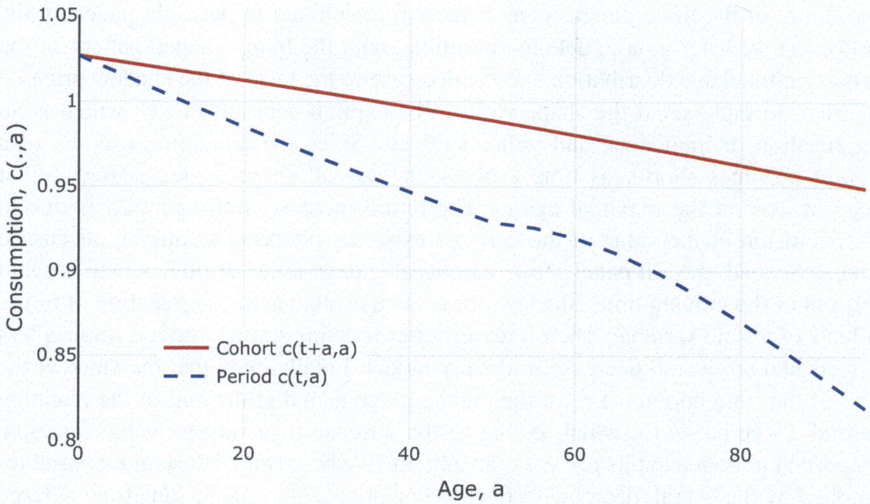


**Fig. 4.3** Optimal consumption along a cohort (red line) and across cohorts (blue dashed line) for a fixed $t$

and vitality rates of individuals.[13] This allows us to directly identify the impact of pollution on the timing of consumption. As pollution increases over time (see Fig. 4.5), the negative effects through all channels become stronger, which is reflected by a decrease in the corresponding negative shadow price $\eta^P$. Moreover, the marginal utility w.r.t. consumption decreases in pollution, as implied by the specific functional form shown in the second term in (4.14). Both effects give an incentive to shift consumption to younger ages and to decrease it continuously over the life-cycle. For the consumption profile across cohorts the explanation is similar, but the profile here combines the consumption values of different cohorts. Earlier-born and thus older cohorts in the cross-section face a lower productivity (lower wage profile) and a higher shadow price $\xi^N(t, a)$, which enters in the definition of $\eta^P$ and is based on a longer remaining time horizon. Therefore, the consumption profile along the time dimension decreases more steeply in the cross-section as compared to the consumption profile along the life-cycle.

As the shadow price of the population plays a decisive role for the development of $\eta^P$ and consequently the optimal consumption allocation, Fig. 4.4 provides more insights into the development of $\xi^N(t, a)$ across time and age. The shadow price is illustrated in two different ways. The left panel shows several cross-sectional age schedules (i.e., vertical lines in the Lexis diagram, Fig. 4.1) of the shadow price for different time periods $t$. These schedules correspond to cross-sectional slices in

[13] No impact of pollution would imply $\kappa(a) = 0$, eliminating the second term in (4.14) as well as $\mu_P = w_P = \nu_p = 0$. Following Eq. (4.10d) this directly leads to $\eta^P = 0$ and consequently $\eta_t^P = 0$.

the figure of the three-dimensional shadow price plotted in the right panel. In the following we offer a more detailed intuition about the hump-shaped pattern of the cross-sectional age distribution and the decrease in the level of the shadow price.

First, to understand the shape we use the explicit solution (4.13), which is an aggregation of immediate and indirect effects. Since the remaining lifetime of a cohort becomes shorter as time evolves, the overall shape is decreasing until it reaches zero at the maximal age $\omega$. The initial increase until age $\approx 25$ is due to the inclusion of the value of the cohort's expected offspring as already discussed above. Second, the left panel shows a noticeable decrease over time, which is again a result of the moving time. Shadow prices always include the aggregation of future effects of a state variable. Therefore, a shorter remaining time horizon implies less effects and an overall decrease in the toy model. Finally, note that the value at the end of the time horizon (i.e., at the maximal age $\omega$ and at the end of the planning period $T$) equals zero, which is due to the absence of a salvage value function. Important to notice in this respect is an anticipative behavior of the control variables implied by the overall decrease of the shadow prices. For a fairly long time horizon that means that the system stabilizes after some transitional initial period before the nearing end of the time horizon leads to a deviation which, in some cases, can be quite counterintuitive. To avoid those effects and caveats, solutions are often plotted on a truncated time horizon.

Note that the right-hand side of Fig. 4.4 also demonstrates that all control, state, and corresponding adjoint variables are derived in the full time-age spectrum, as a PDE. Paths along a cohort or cross cohort (as in Figs. 4.2 and 4.3 or in the left panel of Fig. 4.4) are only slices of the variable in the full time-age domain, but often more suitable to highlight specific effects.

In the discussion of the consumption profile we already mentioned that pollution increases over time. Figures 4.5 and 4.6 present a sensitivity analysis of the pollution effect and compare a high, low, and zero pollution impact. The left panel of Fig. 4.5 shows the pollution over time for the three cases. The gradual increase of pollution over most of the time horizon is due to the increasing population (i.e., more people

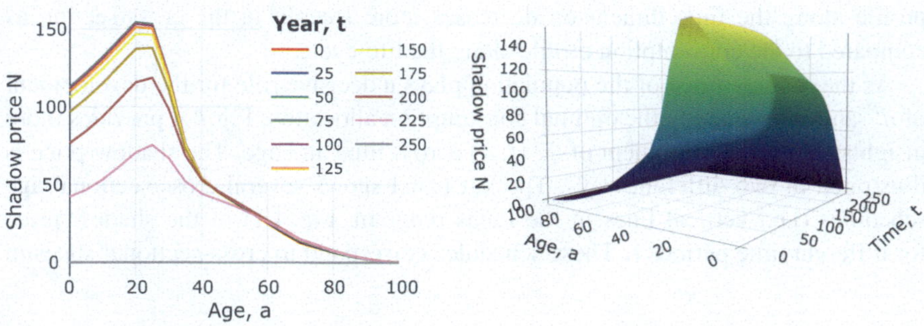


**Fig. 4.4** Adjoint variable of population $\xi^N(t, a)$. The left panel shows a cross-cohort spectrum (i.e., $\xi^N(t, \cdot)$) for specific $t$. The right panel shows the trajectory in the full time-age domain $D$

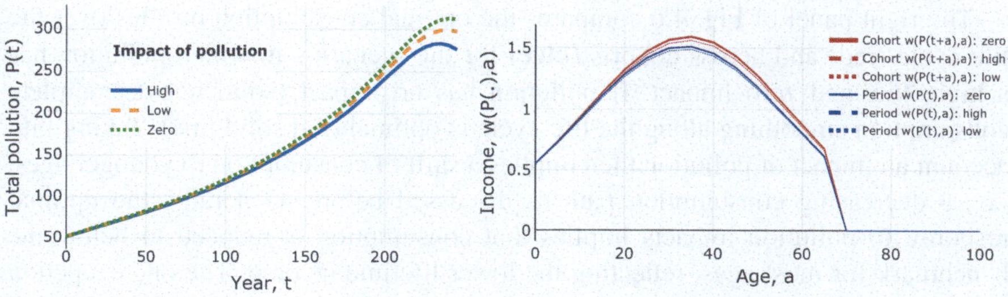


**Fig. 4.5** Impact of pollution. The left panel plots pollution over time with high (blue solid), with low (red dashed), and no (green dotted) effect on the model. The right panel plots the income along a cohort (red lines) and cross cohort for fixed $t$ (blue lines)

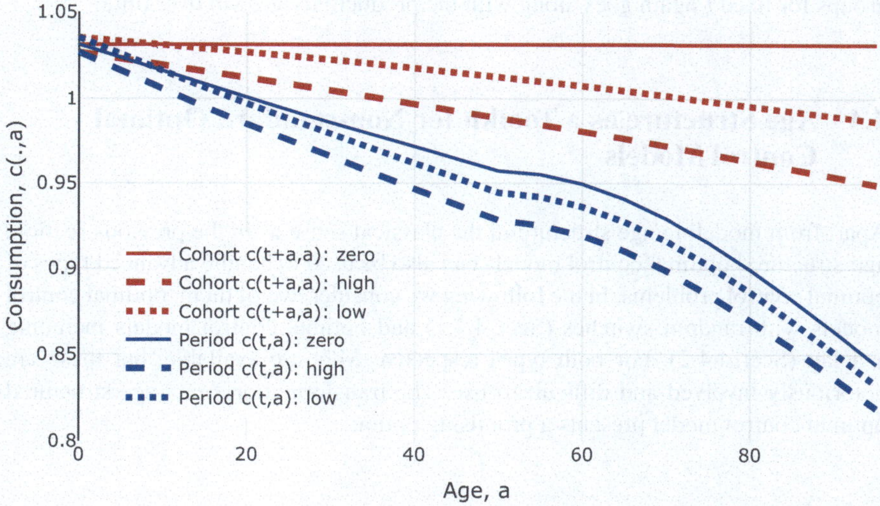


**Fig. 4.6** Impact of pollution on optimal consumption along a cohort (red lines) and cross cohort for a fixed $t$ (blue lines)

with the same wage profile consume more products) and productivity (i.e., higher income)[14]. Thus, these cohorts can only consume less, which also means less air pollution. A comparison between the different schedules in the left panel of Fig. 4.6 shows that a higher impact of pollution on productivity and the vital rates is associated with less aggregate pollution. This reflects the (stronger) effort by the planner towards mitigating pollution by way of reallocating consumption as shown in the right panel of Fig. 4.6.

[14] The decrease at the end of the time horizon is due to the end condition for assets $A(T, a) = 0$. This implies that late-born individuals earn a lower life-cycle income as their most productive age is reached byond the end of the planning period.

The right panel of Fig. 4.6 compares the optimal consumption profiles over the life-cycle (red) and across cohorts (blue) for the scenarios in which pollution has a high, low and zero impact. If pollution has no impact (solid lines), complete consumption smoothing along the life-cycle is optimal (red solid line). Taking into account an impact of pollution then implies a shift of consumption to younger ages, i.e., a decreasing consumption path as discussed before. Generally, the optimal response to pollution impacts implies that consumption is reduced to below the benchmark for most ages, reflecting the lower lifetime income. The one excpetion is the case of low pollution impacts, where consumption is slightly increased above the benchmark for young ages (between 0 and 20). The arguments carry over to the cross-cohort consumption paths (blue lines) where the consumption for cohorts with a younger age is also higher if pollution has an impact. The decrease across the age groups for fixed $t$ again goes along with the productivity growth over time.

## 4.4 Age Structure as a Toolkit for Non-standard Optimal Control Models

Apart from modeling age structure in the classical sense as in the previous section, age-structured optimal control models can also be used to handle advanced types of optimal control problems. In the following we consider two of them: optimal control models with random switches (Sect. 4.4.1) and optimal control models including a delay (Sect. 4.4.2). For both types respective MPs are available, but these are notoriously involved and difficult to use. The transformation to an age-structured optimal control model presents a promising option.

### *4.4.1 Optimal Control Models with Random Switches*

Optimal control models with multiple stages (referred to as multi-stage optimal control models) are well developed if the stage switches are endogenously determined by the decision maker. At the switching point the so-called switching conditions (see Tomiyama, 1985; Tomiyama and Rossana, 1989 or Makris, 2001) have to hold in addition to the standard MP, and a solution can be found by a standard numerical approach.

For stochastic switches the conditions are different and the analysis is more involved. There is no unique second stage, but infinitely many, starting at all possible instants of time. Consequently, the decision maker has to consider all possible switching times (with corresponding stages) and anticipate them in the optimization nexus. She thereby has to consider the effects of her decisions on (i) the level of preparedness for the impacts of a switch and (ii) the likelihood of the switch occurring (as is described by the hazard rate of the switch).

The literature of optimal control models with stochastic switches emerged in the 1970s in environmental economics (see Cropper, 1976; Reed, 1987) and spilled over to economics (Guo et al., 2005), epidemiology (Brock and Xepapadeas, 2020), and other fields. Most of these models use the so-called backward approach, i.e., the deterministic reformulation of the stochastic optimal control model (Boukas et al., 1990), where the value function for the second stage after the switch is derived either analytically or numerically. In the following we introduce the general formulation of an optimal control model with random switching time and present the transformation to an age-structured optimal control model (see Wrzaczek et al., 2020).

Let $x(t)$ and $u(t) \in U$ denote the state and control vectors at $t$ in a standard optimal control model where $F(x(t), u(t), t)$ and $f(x(t), u(t), t)$ are the objective functional and system dynamics. The time horizon is separated into two stages by the switching time $\tau$ (random variable out of the sample space $\Omega = [0, \infty)$, with probability space $(\Omega, \Sigma, \mathbb{P})$), which is stochastic according to the hazard rate $\eta$ that depends on the state and control vectors at $t$, i.e.,

$$\eta(x(t), u(t), t) = \frac{\mathcal{F}'(t)}{1 - \mathcal{F}(t)}, \qquad \mathcal{F}(t) = \mathbb{P}(\tau \leq t). \tag{4.15}$$

At the switch, the model changes disruptively according to three possibilities: (i) change of the objective functional, (ii) change of the system dynamics (including the addition of further or removal of existing state variables), and/or (iii) jump in a state variable. Combining these three possibilities makes it possible to model a broad variety of different effects associated with disruptive regime shifts.

We denote (i) and (ii) by adding subscripts to the corresponding functions; (iii) is modeled via a function $\varphi(x(t), u(t), t)$ that embraces a possible jump. In the case of a switch at $\tau$, the limit of $\varphi$ from the left defines the initial state value of the second stage at $\tau$ (see Eq. (4.17b)).

We adopt the standard notation and introduce the value function of the second stage problem as the salvage value of the first stage and arrive at the following general model[15]

$$\max_{u(t)\in U} \quad \mathbb{E}_{\tau\in[0,\infty)} \left[ \int_0^\tau e^{-\rho t} F_1(x(t), u(t), t) \mathrm{d}t + e^{-\rho\tau} S(x(\tau), \tau) \right] \tag{4.16a}$$

$$\text{s.t. } \dot{x}(t) = f_1(x(t), u(t), t), \quad x(0) = x_0 \tag{4.16b}$$

$$\eta(t) = \eta(x(t), u(t), t), \tag{4.16c}$$

[15] Note the difference in notation in comparison to multi-stage optimal control models (with endogenous switch): In these models the objective function of both stages can be written together as only one switch occurs. In the case of a random switch the control of the second stage will differ for every single switch and therefore depend on the realization of $\tau$. The control variable $u(t)$ cannot therefore be put within one maximization operator.

where

$$S(x(\tau),\tau) := \max_{u(t)\in U} \int_{\tau}^{\infty} e^{-\rho t} F_2(x(t),u(t),t)\mathrm{d}t \tag{4.17a}$$

$$\text{s.t. } \dot{x}(t) = f_2(x(t),u(t),t), \quad x(\tau) = \lim_{t'\nearrow\tau} \varphi(x(t'),u(t'),t'). \tag{4.17b}$$

In general, (4.16) is a stochastic optimal control problem w.r.t. the time horizon. However, by introducing an auxiliary state variable $z_1(t)$ (see Boukas et al., 1990) that evolves according to

$$\dot{z}_1(t) = -\eta(x(t),u(t),t)z_1(t), \qquad z_1(0) = 1, \tag{4.18}$$

it is possible to formulate (4.16) as a deterministic optimal control model. $z_1(t)$ can be interpreted as the probability that the switch has not set in during the interval $[0,t)$ (analogously to a survival probability). Exploiting this deterministic formulation of (4.16) and the auxiliary state variable, Wrzaczek et al. (2020) propose the transformation into an age-structured optimal control problem by considering every possible switching time to generate a new "cohort" and by denoting the corresponding state and control variables (for a second stage initiated by a switch at $t-a$) by $y(t,a)$ and $v(t,a)$.

The full model in age-structured form reads:

$$\max_{u(t)\in U, v(t,a)\in V} \int_0^{\infty} e^{-\rho t} \left[z_1(t)F_1(x(t),u(t),t) + Q(t)\right] \mathrm{d}t \tag{4.19a}$$

$$\text{s.t. } \dot{x}(t) = f_1(x(t),u(t),t), \quad x(0) = x_0 \tag{4.19b}$$

$$\dot{z}_1(t) = -\eta(x(t),u(t),t)z_1(t), \quad z_1(0) = 1 \tag{4.19c}$$

$$\left(\frac{\partial}{\partial t} + \frac{\partial}{\partial a}\right) y(t,a) = f_2(y(t,a),v(t,a),t),$$

$$y(t,0) = \varphi(x(t),u(t),t) \tag{4.19d}$$

$$\left(\frac{\partial}{\partial t} + \frac{\partial}{\partial a}\right) z_2(t,a) = 0, \quad z_2(t,0) = z_1(t)\eta(x(t),u(t),t) \tag{4.19e}$$

$$y(0,a) = z_2(0,a) = 0, a \geq 0 \tag{4.19f}$$

$$Q(t) = \int_0^{\infty} z_2(t,a)F_2(y(t,a),v(t,a),t)\mathrm{d}a. \tag{4.19g}$$

Here, (4.19a) denotes the deterministic formulation of the objective function, (4.19b) and (4.19c) the system dynamics (state variable and probability that the switch has not set in at $t$) during stage 1 (before the switch), and (4.19c) and (4.19d) the system dynamics[16] (state variable and auxiliary state variable $z_2(t, a)$ which is the probability density that the switch happened at $t - a$[17]). The aggregated state $Q(t)$ covers the objective functionals of all possible switches that may have happened before $t$ weighted by the corresponding density.[18]

The advantages of our age-structured formulation in comparison to the backward approach are twofold. Firstly, problem (4.19) can be treated with established numerical solution methods (as briefly discussed in Sect. 4.2) offering a structured (while not always easy) way to solve concrete models. Freiberger (2023) provides a specialized toolbox for the numerical solution of this type of problems implemented in the Julia programming language. In contrast to the age-structured transformation approach the backward approach is often plagued by the curse of dimensionality. As the number of state variables increases, the dimension of the slice manifold along which the value function of stage 2 has to be evaluated increases, too. Consequently, the number of grid points (for which the value function has to be calculated) increases exponentially with the number of state variables.

Secondly, analytical and structural insights are limited in the case of the backward approach, because (and in analogy to a dynamic programming approach) the second stage has to be solved first in order to subsequently obtain a solution for stage 1. In contrast, the age-structured formulation allows to solve stage 1 and stage 2 simultaneously within a single set of optimality conditions and by way of a single (numerical) optimization routine. The solution then reveals the links between the two stages explicitly and allows to characterize the mechanisms of the model in a natural and intuitive way.

The idea of considering optimal control models with a random switch as age-structured optimal control models is still rather new, but has already been used in a number of applications. Kuhn and Wrzaczek (2021) consider a model of rational experimentation with an addictive good where the switch to addiction is modeled as a stochastic shock (that embraces a Skiba point in the second stage). Wrzaczek (2021) includes the risk of catastrophic climate change in an overlapping generations (OLG) model on pollution control. Buratto et al. (2022) consider the development of a vaccine protecting against COVID-19 as a positive stochastic shock and analyze anticipative behavior in the stage without vaccination and the optimal adaptation of pandemic countermeasures shortly after the start of the vaccination rollout. Freiberger et al. (2023) analyze the optimal patterns of consumption and health-care utilization over the individual life-cycle in view of

---

[16] Note that the limit can be neglected in (4.19d) as the notation of the state variable changes from $x(\tau)$ to $y(\tau, 0)$ at the switch at $\tau$.

[17] As argued in Wrzaczek et al. (2020), the auxiliary state $z_2(t, a)$ avoids having to deal with a time-lag in (4.19g).

[18] Note that the time horizon here is infinite although we introduced the MP only for age-structured optimal control problems in finite time in Theorem 4.1. However, the present model (4.19) is a special case as the different "cohorts" do not interact with each other and their aggregation only enters the objective function. It can be shown for this case that the MP also holds for problems with infinite time horizon.

large shocks to health. This paper exploits the advantages of the age-structured approach and carefully disentangles different channels of the optimal allocation of preventive, acute, and chronic care. Buratto et al. (2024) consider an advertising model with an abrupt change of the production costs, where the random switch depends positively on demand.

### 4.4.2 *Optimal Control Models with Time-Lags or Delays*

Optimal control models with time-lags or delays are other advanced extensions of standard optimal control theory. These models can be divided into two classes: (i) Models with continuous time-lags correspond to a class of models where the system dynamics and/or the objective function at $t$ depends on the previous path—or a part of it—of the control or state variable. (ii) Models with delay include the dependence of the system dynamics and/or the objective function on the state and control variables at one specific past point in time $t - \tau$, where $\tau > 0$ denotes the delay.

Formal proofs for (i) can be found in Bate (1969) or Vinokurov (1969), sufficiency conditions have been shown by Sethi (1974). For a textbook representation we refer to Feichtinger and Hartl (1988) and applications can be found, e.g., in Sethi and McGuire (1977), Arthur and McNicoll (1977), Hartl and Sethi (1984), Caulkins et al. (2010), or Boucekkine et al. (2004). The first proof for (ii) goes back to Kharatishvili (1961), which has been extended by Halanay (1968) to the case of multiple delays (equal for state and control). Göllmann et al. (2009) add mixed control-state constraints to the problem.

The literature on applications of optimal control models with time-lags is still relatively scarce in spite of the many new developments, extensions, and applications of optimal control theory over the past decades. The reason appears to be twofold: First, although the theoretical contributions provide necessary optimality conditions, the theory is advanced, and it is more difficult to obtain analytical as well as numerical results. Second, a time-lag in state and control variables is often modeled as an aggregated state variable that approximates a certain effect at $t$. Given the complex nature of optimal control problems in general, the second argument then usually implies the use of an approximation to guarantee tractability of the model. This is striking, given the importance of applications in which the time-lag is crucial, e.g., the construction of a dam, which requires several years of planning and construction, followed by multiple years of rising water levels until the benefits of the investment can be realized (hydropower, flood control, etc.).

Although models with a continuous time-lag and models with delay are treated separately in the literature, we will work with a continuous time-lag and argue why it can also be used for a delay, at least as an arbitrarily close approximation. Let us thus consider the following optimal control model with continuous delay:

$$\max_{u(t)\in U} \int_0^T e^{-\rho t} F(x(t), u(t), \phi(t), t)\mathrm{d}t + e^{-\rho T} S(x(T), \phi(T), T) \tag{4.20a}$$

$$\text{s.t. } \dot{x}(t) = f(x(t), u(t), \phi(t), t), \quad x(0) = x_0 \tag{4.20b}$$

$$\phi(t) = \int_{-\infty}^{t} g(x(s), u(s), s, t)\mathrm{d}s \tag{4.20c}$$

$$x(t) = \tilde{x}(t), \quad u(t) = \tilde{u}(t), \quad \text{for } t < 0, \tag{4.20d}$$

where control and state variables and functions are denoted in the same way as above. The function $\phi(t)$ is now governing the continuous time-lag by aggregating the density of the effects of all past (i.e., in general for $s \in (-\infty, t)$) control and state variables. Thus, $\phi(t)$ enters the objective functional, the salvage value function, and the system dynamics. To capture past effects reaching back before the start of the planning period (i.e., $t < 0$), condition (4.20d) adds the part of the trajectory before the planning horizon.

In case the time-lag is a fixed delay, $\phi(t)$ does not depend on the entire history of the control and state variables, but on a specific time alone. This dependence is described by the function $\tilde{g}$:

$$\phi(t) = \tilde{g}(x(t-\tau), u(t-\tau), t-\tau, t), \; x(t) = \tilde{x}(t), u(t) = \tilde{u}(t), \; \text{for } t \in [-\tau, 0). \tag{4.21}$$

This definition can be extended straightforwardly to cover the case of multiple fixed delays. Note that (4.21) can be transformed to fit into framework (4.20) by employing the Dirac delta function $\delta(t)$,

$$\begin{aligned}\phi(t) &= \tilde{g}(x(t-\tau), u(t-\tau), t-\tau, t)\\ &= \int_0^{\infty} \delta(\tau - s')\tilde{g}(x(t-s'), u(t-s'), t-s', t)\mathrm{d}s'\\ &= \int_{-\infty}^{t} \underbrace{\delta(s-(t-\tau))\tilde{g}(x(s), u(s), s, t)}_{=:g(x(s),u(s),s,t)}\mathrm{d}s\end{aligned}$$

such that state and control enter $\phi(t)$ only by the delay.[19]

In order to formulate (4.20) as an age-structured optimal control problem, we consider two new auxiliary (age-structured) state variables emerging at every $t$

[19] Note that this argument is not entirely correct in mathematical terms as the delta distribution is not integrable in a Riemann and Lebesgue sense. However, the delta function can be defined as the limit of a series of probability density functions $(\delta_k)_{k\in\mathbb{N}}$ (imagine a series of normal distributions with mean 0 and a variance converging to 0). Each function in the series is integrable and can be used to approximate the Dirac function to an arbitrary degree of freedom, which translates to an arbitrarily close approximation of the case of fixed delay, based on a probability density function.

with zero dynamics at the corresponding boundary condition. One of the "cohorts" accounts for the control and the other for the state variable at $t$. The initial distributions for these auxiliary states cover the state and control trajectories before the planning horizon (denoted by $\tilde{x}(t)$ and $\tilde{u}(t)$ in (4.20)). The function $\phi(t)$ in turn is represented by an aggregate state variable, as standard in the general model (4.2).

The full model in age-structured form then reads:

$$\max_{u(t)\in U} \quad \int_0^T e^{-\rho t} F(x(t), u(t), \phi(t), t)\mathrm{d}t + e^{-\rho T} S(x(T), \phi(T), T) \tag{4.22a}$$

$$\text{s.t.} \quad \dot{x}(t) = f(x(t), u(t), \phi(t), t), \qquad x(0) = x_0 \tag{4.22b}$$

$$\left(\frac{\partial}{\partial t} + \frac{\partial}{\partial a}\right) y^1(t, a) = 0, \qquad y^1(t, 0) = x(t) \tag{4.22c}$$

$$\left(\frac{\partial}{\partial t} + \frac{\partial}{\partial a}\right) y^2(t, a) = 0, \qquad y^2(t, 0) = u(t) \tag{4.22d}$$

$$\phi(t) = \int_0^\infty g(y^1(t, a), y^2(t, a), t - a, t)\mathrm{d}a \tag{4.22e}$$

$$y^1(0, a) = \tilde{x}(a), \qquad \text{for } a \geq 0 \tag{4.22f}$$

$$y^2(0, a) = \tilde{u}(a), \qquad \text{for } a \geq 0. \tag{4.22g}$$

To the best of our knowledge, this transformation has not been investigated (at least) explicitly in the literature. However, as discussed above, the analytical treatment and the numerical methods are developed for such a model. The advantages of these approaches are similar to the ones of the age-structured formulation of optimal control models with random switches.

## 4.5 Discussion and Conclusions

Age-structured optimal control models are important frameworks to take into account the interplay between cohort and period effects in many applications in economics, environmental science, epidemiology, and many more disciplines. Setting up a generic age-structured optimal control model, we have shown how such models allow for aggregated and distributed state variables as well as concentrated and age- and time-dependent control variables. We introduced the analytical results of the age-structured maximum principle and sketched the numerical solutions of these models.

Based on a toy model of a social planner aiming to reduce aggregate pollution generated by consumption when maximizing the discounted stream of future utility for a society, we demonstrate the application of the age-structured optimal control model. Within our toy model we can show how inter- and intra-cohort pollution effects interact and how the social planner can internalize these effects. Considering the present value of an additional consumer, i.e., the shadow value of an individual, we show how cohort redistribution effects and population dynamic effects are intertwined with optimal redistributions across the life-cycle.

We end our review by showing two examples where age-structured optimal control models can be applied to solve non-standard optimal control models. First, we introduce control models with random switches. By defining every possible switching time to generate a new cohort, the toolkit of age-structured optimal control models allows to apply established numerical solution methods. Second, we introduce a rather novel transformation of an optimal control model with time-lags or delays into an age-structured optimal control model.

## References


Aguilar-Gomez, S., Dwyer, H., Graff Zivin, J., & Neidell, M. (2022). This is air: The "nonhealth" effects of air pollution. *Annual Review of Resource Economics, 14*(1), 403–425. https://doi.org/10.1146/annurev-resource-111820-021816.

Almeder, C., Caulkins, J. P., Feichtinger, G., & Tragler, G. (2004). An age-structured single-state drug initiation model—cycles of drug epidemics and optimal prevention programs. *Socio-Economic Planning Sciences, 38*(1), 91–109.

Almetwally, A. A., Bin-Jumah, M., & Allam, A. A. (2020). Ambient air pollution and its influence on human health and welfare: an overview. *Environmental Science and Pollution Research, 27*(20), 24815–24830.

Arthur, W. B., & McNicoll, G. (1977). Optimal time paths with age-dependence: A theory of population policy. *The Review of Economic Studies, 44*(1), 111.

Augeraud-Véron, E., Boucekkine, R., & Veliov, V. M. (2019). Distributed optimal control models in environmental economics: A review. *Mathematical Modelling of Natural Phenomena, 14*(1), 106.

Bate, R. R. (1969). The optimal control of systems with transport lag. In C. T. Leondes (Ed.), *Advances in control systems* (Vol. 7, pp. 165–224). Elsevier.

Bonnans, J. F. (2013). The shooting approach to optimal control problems. *IFAC Proceedings Volumes, 46*(11), 281–292.

Boucekkine, R., Croix, D. D. L., & Licandro, O. (2004). Modelling vintage structures with DDEs: Principles and applications. *Mathematical Population Studies, 11*(3–4), 151–179.

Boukas, E. K., Haurie, A., & Michel, P. (1990). An optimal control problem with a random stopping time. *Journal of Optimization Theory and Applications, 64*(3), 471–480.

Brock, W., & Xepapadeas, A. (2020). The economy, climate change and infectious diseases: Links and policy implications. *Environmental and Resource Economics, 76*(4), 811–824.

Brogan, W. L. (1968). Optimal control theory applied to systems described by partial differential equations. In *Advances in control systems* (Vol. 6, pp. 221–316). Elsevier.

Brokate, M. (1985). Pontryagin's principle for control problems in age-dependent population dynamics. *Journal of Mathematical Biology, 23*(1), 75–101.

Buratto, A., Cesaretto, R., & Grosset, L. (2020). Infinite-horizon linear-quadratic age-structured optimal control problem. *Applied Mathematical Sciences, 14*(9), 409–416.

Buratto, A., Muttoni, M., Wrzaczek, S., & Freiberger, M. (2022). Should the COVID-19 lockdown be relaxed or intensified in case a vaccine becomes available? *PLOS ONE, 17*(9), e0273557.

Buratto, A., Grosset, L., Muttoni, M., & Viscolani, B. (2025). The cost of myopia with respect to a switching time in an advertising model. In *The unaffordable price of static decision-making models*. Springer.

Caulkins, J. P., Hartl, R. F., & Kort, P. M. (2010). Delay equivalence in capital accumulation models. *Journal of Mathematical Economics, 46*(6), 1243–1246.

Chan, W., & Guo, B. Z. (1990). Optimal birth control of population dynamics. II. Problems with free final time, phase constraints, and mini-max costs. *Journal of Mathematical Analysis and Applications, 146*(2), 523–539.

Chan, W., & Guo, G. B. (1989). Optimal birth control of population dynamics. *Journal of Mathematical Analysis and Applications, 144*(2), 532–552.

Coale, A. J. (1957). A new method for calculating Lotka's r–the intrinsic rate of growth in a stable population. *Population Studies, 11*(1), 92–94. Publisher: [Population Investigation Committee, Taylor & Francis, Ltd.]

Conforti, A., Mascia, M., Cioffi, G., De Angelis, C., Coppola, G., De Rosa, P., Pivonello, R., Alviggi, C., & De Placido, G. (2018). Air pollution and female fertility: a systematic review of literature. *Reproductive Biology and Endocrinology: RB&E, 16*(1), 117.

Cropper, M. L. (1976). Regulating activities with catastrophic environmental effects. *Journal of Environmental Economics and Management, 3*(1), 1–15.

Derzko, N., Sethi, S. P., & Thompson, G. L. (1980). Distributed parameter systems approach to the optimal cattle ranching problem. *Optimal Control Applications and Methods, 1*(1), 3–10.

Feichtinger, G., & Hartl, R. F. (1988). *Optimale Kontrolle Ökonomischer Prozesse: Anwendungen des Maximumprinzips in den Wirtschaftswissenschaften*. Walter de Gruyter.

Feichtinger, G., Hartl, R. F., Kort, P. M., & Veliov, V. M. (2006). Anticipation effects of technological progress on capital accumulation: a vintage capital approach. *Journal of Economic Theory, 126*(1), 143–164.

Feichtinger, G., Krasovskii, A. A., Prskawetz, A., & Veliov, V. M. (2012). Optimal age-specific election policies in two-level organizations with fixed size. *Central European Journal of Operations Research, 20*(4), 649–677.

Feichtinger, G., Kuhn, D. R., Prskawetz, A., & Wrzaczek, S. (2011). The reproductive value as part of the shadow price of population. *Demographic Research, 24*, 709–718.

Feichtinger, G., Prskawetz, A., & Veliov, V. M. (2004). Age-structured optimal control in population economics. *Theoretical Population Biology, 65*(4), 373–387.

Feichtinger, G., Tragler, G., & Veliov, V. M. (2003). Optimality conditions for age-structured control systems. *Journal of Mathematical Analysis and Applications, 288*(1), 47–68.

Feichtinger, G., & Wrzaczek, S. (2024a). The optimal momentum of population growth and decline. *Theoretical Population Biology, 155*, 51–66.

Feichtinger, G., & Wrzaczek, S. (2024b). The optimal transition to a stationary population for concentrated vitality rates. *Demographic Research, 50*, 171–184.

Freiberger, M. (2023). Two-Stage-Optimal-Control (v0.1.0). Zenodo. https://doi.org/10.5281/zenodo.7772076

Freiberger, M., Kuhn, M., & Wrzaczek, S. (2023). Modeling health shocks. *IIASA Working Paper WP-23-002*.

Gao, X., Song, R., & Timmins, C. (2022). The fertility consequences of air pollution in China. *National Bureau of Economic Research 30165*.

Göllmann, L., Kern, D., & Maurer, H. (2009). Optimal control problems with delays in state and control variables subject to mixed control-state constraints. *Optimal Control Applications and Methods, 30*(4), 341–365.

Greenhalgh, D. (1988). Some results on optimal control applied to epidemics. *Mathematical Biosciences, 88*(2), 125–158.

Guo, X., Miao, J., & Morellec, E. (2005). Irreversible investment with regime shifts. *Journal of Economic Theory, 122*(1), 37–59.

Gurtin, M. E., & Murphy, L. F. (1981a). On the optimal harvesting of age-structured populations: Some simple models. *Mathematical Biosciences, 55*(1), 115–136.

Gurtin, M. E., & Murphy, L. F. (1981b). On the optimal harvesting of persistent age-structured populations. *Journal of Mathematical Biology, 13*(2), 131–148.

Halanay, A. (1968). Optimal controls for systems with time lag. *SIAM Journal on Control, 6*(2), 215–234. Publisher: Society for Industrial and Applied Mathematics.

Hartl, R. F., Kort, P. M., & Wrzaczek, S. (2023). Reputation or warranty, what is more effective against planned obsolescence? *International Journal of Production Research, 61*(3), 939–954.

Hartl, R. F., & Sethi, S. P. (1984). Optimal control of a class of systems with continuous lags: Dynamic programming approach and economic interpretations. *Journal of Optimization Theory and Applications, 43*(1), 73–88.

Hethcote, H. W. (1988). Optimal ages of vaccination for measles. *Mathematical Biosciences, 89*(1), 29–52.

Horiuchi, S., Ouellette, N., Cheung, S. L. K., & Robine, J.-M. (2014). Modal age at death: lifespan indicator in the era of longevity extension. *Vienna Yearbook of Population Research Volume, 11*, 37–69.

Jurewicz, J., Dziewirska, E., Radwan, M., & Hanke, W. (2018). Air pollution from natural and anthropic sources and male fertility. *Reproductive Biology and Endocrinology: RB&E, 16*(1), 109.

Keyfitz, B., & Keyfitz, N. (1997). The McKendrick partial differential equation and its uses in epidemiology and population study. *Mathematical and Computer Modelling, 26*(6), 1–9.

Kharatishvili, B. (1961). *A maximum principle in external problems with delays*. Doklady Akademii Nauk USSR.

Kuhn, M., & Wrzaczek, S. (2021). Rationally risking addiction: A two-stage approach. In J. L. Haunschmied, R. M. Kovacevic, W. Semmler, & V. M. Veliov (Eds.) *Dynamic economic problems with regime switches* (pp. 85–110). Springer International Publishing.

Kuhn, M., Wrzaczek, S., & Oeppen, J. (2010). Recognizing progeny in the value of life. *Economics Letters, 107*(1), 17–21.

Kuhn, M., Wrzaczek, S., Prskawetz, A., & Feichtinger, G. (2011). Externalities in a life cycle model with endogenous survival. *Journal of Mathematical Economics, 47*(4–5), 627–641.

Leitmann, G., & Stalford, H. (1971). A sufficiency theorem for optimal control. *Journal of Optimization Theory and Applications, 8*(3), 169–174.

Makris, M. (2001). Necessary conditions for infinite-horizon discounted two-stage optimal control problems. *Journal of Economic Dynamics and Control, 25*(12), 1935–1950.

Medhin, N. G. (1992). Optimal harvesting in age-structured populations. *Journal of Optimization Theory and Applications, 74*(3), 413–423.

Neidell, M. P. (2023). Air pollution and worker productivity. *IZA World of Labor*.

Peszko, G., Amann, M., Awe, Y., Kleiman, G., & Rabie, T. S. (2023). *Air pollution and climate change: From co-benefits to coherent policies*. The World Bank.

Prskawetz, A., Tsachev, T., & Veliov, V. M. (2012). Optimal education in an age-structured model under changing labor demand and supply. *Macroeconomic Dynamics, 16*(2), 159–183.

Prskawetz, A., & Veliov, V. M. (2007). Age-specific dynamic labor demand and human capital investment. *Journal of Economic Dynamics and Control 31*(12), 3741–3777.

Rao, N. D., Kiesewetter, G., Min, J., Pachauri, S., & Wagner, F. (2021). Household contributions to and impacts from air pollution in India. *Nature Sustainability, 4*(10), 859–867. Number: 10 Publisher: Nature Publishing Group.

Reed, W. J. (1987). Protecting a forest against fire: Optimal protection patterns and harvest policies. *Natural Resource Modeling, 2*(1), 23–53. https://onlinelibrary.wiley.com/doi/pdf/10.1111/j.1939-7445.1987.tb00025.x.

Seierstad, A., & Sydsaeter, K. (1977). Sufficient conditions in optimal control theory. *International Economic Review 18*(2), 367–391. Publisher: [Economics Department of the University of Pennsylvania, Wiley, Institute of Social and Economic Research, Osaka University].

Sethi, S. P. (1974). Sufficient conditions for the optimal control of a class of systems with continuous lags. *Journal of Optimization Theory and Applications, 13*(5), 545–552.

Sethi, S. P., & McGuire, T. W. (1977). Optimal skill mix: An application of the maximum principle for systems with retarded controls. *Journal of Optimization Theory and Applications, 23*(2), 245–275.

Shi, D., & Yu, H. (2020). Reevaluating the subjective welfare loss of air pollution. *Journal of Cleaner Production, 257*, 120445.

Skritek, B., & Veliov, V. M. (2015). On the Infinite-Horizon optimal control of age-structured systems. *Journal of Optimization Theory and Applications, 167*(1), 243–271.

Tomiyama, K. (1985). Two-stage optimal control problems and optimality conditions. *Journal of Economic Dynamics and Control, 9*(3), 317–337.

Tomiyama, K., & Rossana, R. J. (1989). Two-stage optimal control problems with an explicit switch point dependence: Optimality criteria and an example of delivery lags and investment. *Journal of Economic Dynamics and Control 13*(3), 319–337.

Veliov, V. M. (2003). Newton's method for problems of optimal control of heterogeneous systems. *Optimization Methods and Software, 18*(6), 689–703. Publisher: Taylor & Francis _eprint: https://doi.org/10.1080/1055678031000163975З.

Veliov, V. M. (2008). Optimal control of heterogeneous systems: Basic theory. *Journal of Mathematical Analysis and Applications, 346*(1), 227–242.

Vinokurov, V. R. (1969). Optimal control of processes described by integral equations, III. *SIAM Journal on Control, 7*(2), 346–355. Publisher: Society for Industrial and Applied Mathematics.

Wang, P. (1964). Control of distributed parameter systems. In *Advances in control systems* (Vol. 1, pp. 75–172). Elsevier.

Wrzaczek, S. (2021). An OLG differential game of pollution control with the risk of a catastrophic climate change. *International Game Theory Review, 23*(04), 2250002.

Wrzaczek, S., Kuhn, M., & Frankovic, I. (2020). Using age structure for a multi-stage optimal control model with random switching time. *Journal of Optimization Theory and Applications, 184*(3), 1065–1082.

Wrzaczek, S., Kuhn, M., Prskawetz, A., & Feichtinger, G. (2010). The reproductive value in distributed optimal control models. *Theoretical Population Biology, 77*(3), 164–170.

Zachmanoglou, E. C., & Thoe, D. W. (1986). *Introduction to partial differential equations with applications*. Dover Publications.

Zhao, H., Geng, G., Zhang, Q., Davis, S. J., Li, X., Liu, Y., Peng, L., Li, M., Zheng, B., Huo, H., Zhang, L., Henze, D. K., Mi, Z., Liu, Z., Guan, D., & He, K. (2019). Inequality of household consumption and air pollution-related deaths in China. *Nature Communications, 10*(1), 4337.

# Chapter 5
# A Vindication of Open-Loop Equilibria in Differential Games

**Luca Lambertini**

**Abstract** This chapter assesses the properties of open-loop equilibria to clarify their relevance. To this aim, it illustrates the classes of differential games which yield degenerate feedback strategies and equilibria under open-loop information. Then, it discusses the nature of the open-loop solutions of games in which such solutions are strongly time consistent, accounting also for the normative prescriptions one can draw on their basis.

**Keywords** Open-loop equilibrium · Feedback equilibrium · Strong time consistency · Perfect games

**JEL Codes** C61, C73

## 5.1 Introduction

How should we interpret the open-loop solution of a differential game? Or, what are exactly the implications of adopting open-loop information, instead of a closed-loop or feedback one, taking into account that, in general, "closed-loop" and "feedback" are not interchangeable attributes?

Quite often, open-loop solutions are perceived and labelled as *quasi-static*, if not literally *static*. This is a prelude to a view according to which open-loop information adds very little (if anything at all) to what one may learn from the solution of the static version of the same game. Moreover, a frequent additional critique is that open-loop equilibria are usually not feedback ones as they are determined on the basis of initial and transversality conditions and time, disregarding the impact of states on control. At first glance, one could be tempted to share these considerations.

L. Lambertini (✉)
Department of Economics and Alma Climate Research Institute, University of Bologna, Bologna, Italy
e-mail: luca.lambertini@unibo.it

 
F. El Ouardighi, G. Feichtinger (eds.), *The Unaffordable Price of Static Decision-making Models*, International Series in Operations Research & Management Science 365, https://doi.org/10.1007/978-3-031-88638-6_5

Yet, my intent is to show that it would not be a sound choice in many relevant cases, and why. Indeed, the main aims of the ensuing analysis are to offer the essential elements which are necessary to fully appreciate the nature of open-loop strategies and equilibria, to motivate the search for degenerate feedback equilibria attained under open-loop information and, finally, to illustrate in explicit terms the value thereof.

The point of departure concerns the illustration of the concept of *perfect game*, which is the label of a quite large class of games whose open-loop solutions are degenerate feedback ones. The second step is the brief summary of the properties of open-loop and feedback solutions of a famous oligopoly game which, being fully solvable, allows one to characterise analytically any equilibrium engendered under any information structure. In particular, there clearly emerges that, in general, open-loop equilibria replicate static ones only in correspondence of specific limits on key parameters (typically, the discount rate). This exercise is needed to appreciate what follows, namely, a short survey of several perfect games delivering feedback equilibria under open-loop information. This serves the purpose of illustrating the desirability and ductility of such games along both the positive and the normative dimensions. Last but not least—and without spoiling the essence of it—the bottom line of the whole discussion will be that one should welcome the arising of feedback strategies in open-loop games, as these are the only ones validating systematically their counterparts, arising from the analogous static games.

## 5.2 On Strongly Time Consistent Open-Loop Equilibria

To begin with, a few words about terminology. A feedback strategy (and the resulting equilibrium) can be equivalently defined as strongly time consistent, Markov-perfect, or subgame perfect (although one should better resort to this last definition in the domain of static multistage games). The second step consists of a few short considerations about the nature of closed-loop games and their solutions, which I won't dwell upon any more in the remainder. Every feedback strategy is a closed-loop one by construction, while the opposite is not true: in general, a closed-loop strategy is Markovian, but not necessarily perfectly so. To grasp the essence of this fact, it suffices to think of the memoryless closed-loop solution of a differential game defined in Hamiltonian form, with state-control loops appearing in the system of costate equations. If the resulting strategies do not coincide with those delivered by the solution of the corresponding Hamilton-Jacobi-Bellman (HJB) equation, then the nature of the closed-loop solution is Markovian insofar as it features the loops, but not Markov-perfect or strongly time consistent.

Now I may focus specifically upon the open-loop solution, which, in most cases, is evidently not Markov-perfect. However, the possibility for degenerate feedback strategies to arise under open-loop information is one of the cornerstones of the long-lasting debate about the attainment of strong time consistency (or subgame perfectness) in differential games (cf. Dockner et al. 2000, ch. 7). Indeed, any differential game exhibiting this property belongs to the class of the so-called perfect games (Mehlmann, 1988; Cellini et al., 2005), and the stream of research aimed at identifying perfect games and the properties delivering subgame perfectness is a long-standing one.

The first class that has been identified is that of *trilinear games* (Clemhout and Wan, 1974, 1979): a game is trilinear if and only if the state variables do not enter the set of costate equations and optimal controls are independent of states. The second class is that gathering *linear state games*, i.e., those linear in all of the state variables involved (it is worth stressing that this implies that also linear-quadratic games can be state-linear).[1] The third class includes *exponential games*, which can be shown to correspond to linear state ones through a transformation of variables (Reinganum, 1982a,b).

Later contributions have focused on specific properties of the first order conditions on controls, rather than the structure of the game or the objective functionals. To illustrate the matter, consider a differential game involving $N = 1, 2, \ldots n$ players and define the state, costate and control vectors as $\mathbf{x} = \left(x_{1,}x_2, \ldots x_n\right)$, $\boldsymbol{\lambda} = \left(\lambda_{11,}\lambda_{12}, \ldots \lambda_{nn}\right)$ and $\mathbf{u} = \left(x_{1,}x_2, \ldots x_n\right)$. Then let $\pi_i\left(\mathbf{x}, \mathbf{u}\right)$ and $\mathcal{H}_i\left(\mathbf{x}, \mathbf{u}\right)$ be, respectively, the instantaneous payoff and Hamiltonian function of player $i$.

Dockner et al. (1985) have shown that an additional class of perfect games is that in which the following conditions hold:

$$\left.\frac{\partial^2 \mathcal{H}_i}{\partial u_i \partial x_j}\right|_{\frac{\partial \mathcal{H}_i}{\partial u_i}=0} = \left.\frac{\partial^2 \mathcal{H}_i}{\partial u_i \partial x_j}\right|_{u_i^*} = 0 \tag{5.1}$$

$$\frac{\partial^2 \mathcal{H}_i}{\partial x_j^2} = 0 \tag{5.2}$$

for all $i$ and $j$. This class gathers the so-called state-separable games. The first condition requires that the maximised Hamiltonian be independent of any state variable; the second condition requires the Hamiltonian to be state-linear. Hence, one may appreciate that any trilinear or linear state game is necessarily a state-separable one as well, while the opposite is not true.

To complete the picture, one has to add the class of *state-redundant games* (Fershtman, 1987). To save upon notation, think for a moment to the case of a single state $x$, whereby the vector of costates is $\boldsymbol{\lambda} = \left(\lambda_{1,}\lambda_2, \ldots \lambda_n\right)$. A state-redundant game is identified by

$$\left.\frac{\partial^2 \mathcal{H}_i}{\partial u_i \partial x}\right|_{\lambda_i=\lambda_i^*} = 0 \tag{5.3}$$

[1] A well-known example of a differential game whose structure is simultaneously linear-quadratic and state-linear is the unregulated version of the Cournot game with polluting emissions in Benchekroun and Long (1998).

which amounts to requiring the first order condition on $u_i$ to be independent of the state in correspondence of the optimal value of the costate variable. In this case, $\lambda_i^*$ is indeed a proper shadow price, as it coincides with the partial derivative of the value function w.r.t. $x$ in the corresponding HJB equation. We also know from Mehlmann and Willing (1983) that any perfect game is or can be reformulated as a state-redundant game.

From a purely technical standpoint, the most relevant implication of the above discussion is the following: any perfect game admits a representation in the form of a HJB equation delivering a set of Riccati equations whose solutions include the open-loop strategy yielded by the corresponding Hamiltonian representation of the same game. The value added of games featuring this property is assessed in the remainder of the paper, together with an appraisal of the open-loop solution in itself, even when—as it happens in the vast majority of cases—it does not deliver a degenerate feedback control.

## 5.3 Open-Loop vs Feedback Solutions

As a point of departure, I will discuss the difference between open-loop and feedback strategies in a well-known game in which open-loop ones are not Markov-perfect. This is the sticky price game dating back to Simaan and Takayama (1976, 1978), subsequently developed by Fershtman and Kamien (1987) and Tsutsui and Mino (1990), among others.

The game takes place in continuous time over an infinite horizon, with firms 1 and 2 selling a homogeneous good and competing à la Cournot. The common state equation describes the dynamics of price:

$$\dot{p} = s\left[\widehat{p} - p(t)\right] \tag{5.4}$$

where $\widehat{p} = a - q_1(t) - q_2$ is the notional demand function and parameter $s \in [0, \infty)$ measures the time-invariant speed of adjustment of the sticky price $p(t)$. At any time $t$, firms use the same technology, summarised by the cost function $C_i(t) = cq_i(t) + q_i(t)/2$, so that the individual instantaneous profit function is $\pi_i(t) = [p(t) - c - q_i(t)/2]\, q_i(t)$. Assuming a common and constant discount rate $\rho \in [0, \infty)$, firm $i$ has to solve the following problem:

$$\max_{q_i(t)} \int_0^\infty \pi_i(t)\, e^{-\rho t} dt$$

s.t. (5.4) and the initial condition $p(0) = p_0 > 0$.

### 5.3.1 The Open-Loop Game

The Hamiltonian function of firm $i$ is

$$\mathcal{H}_i\left(p(t), q_i(t), \lambda(t)\right) = e^{-\rho t}\left\{\left[p(t) - c - \frac{q_i(t)}{2}\right]q_i(t) + \lambda_i(t)s\left[\widehat{p} - p(t)\right]\right\} \tag{5.5}$$

in which $\lambda_i(t)$ is the capitalised costate variable. The necessary conditions are

$$\frac{\partial \mathcal{H}_i(\cdot)}{\partial q_i(t)} = p(t) - c - q_i(t) - \lambda_i(t)s = 0 \tag{5.6}$$

and

$$-\frac{\partial \mathcal{H}_i(\cdot)}{\partial p(t)} = \dot{\lambda}_i(t) - \rho\lambda_i(t) \Rightarrow \dot{\lambda}_i(t) = \lambda_i(t)(s+\rho) - q_i(t) \tag{5.7}$$

while the transversality condition is $\lim_{t\to\infty} e^{-\rho t}\lambda(t)\, p(t) = 0$. I will omit most of the details, which can be found in Fershtman and Kamien (1987), Dockner et al. (2000), Cellini and Lambertini (2004) and Lambertini (2018). What matters is that the resulting control equation is[2]

$$\dot{q} = s(a - c - 2p) - \rho(p - c - q) \tag{5.8}$$

which, imposing stationarity, delivers the equilibrium strategy under open-loop information:

$$q^{OL}(p) = \frac{p(2s+\rho) - as - c(s+\rho)}{\rho} \tag{5.9}$$

which is monotonically increasing in the current price in the entire admissible parameter space. It is the case of adding that looking at (5.9) or any open-loop strategy being a function of the state variable(s)—which holds whenever the strategy at hand is not a degenerate feedback control—one could say that the open-loop strategy has a quasi-closed-loop nature (while the opposite is not true).[3] Plugging $q^{OL}(p)$ into (5.4) and imposing stationarity, one obtains the steady-state price $p^{OL} = [a(2s+\rho) + 2c(s+\rho)]/(4s+3\rho) > c$, whereby the steady-state output is $q^{OL} = (a-c)(s+\rho)/(4s+3\rho)$.

---

[2] In the remainder of the exposition of the sticky price game, I will omit the explicit indication of the time argument to save upon notation. Likewise, for the same reason, I will also leave aside the discussion of the conditions ensuring the positivity of optimal output strategies.

[3] The closed-loop memoryless equilibrium is illustrated in Cellini and Lambertini (2004) with homogeneous goods, and in Cellini and Lambertini (2007) with differentiated goods.

Considering the aims of the present analysis, the most relevant properties of the open-loop solution are to be identified in the following limits:

$$\lim_{\rho \to 0} q^{OL} = \lim_{s \to \infty} q^{OL} = \frac{a-c}{4} = q^{CN} \tag{5.10}$$

$$\lim_{\rho \to \infty} q^{OL} = \lim_{s \to 0} q^{OL} = \frac{a-c}{3} = q^{pc} \tag{5.11}$$

That is, according to (5.10), if either the adjustment speed of market price is infinitely high or firms do not discount at all future profits, then the open-loop equilibrium reproduces the static Cournot-Nash one. In particular, the first case means that firms are playing along the 'notional' market demand from the very beginning and remain there forever. Otherwise, for all $s, \rho \in (0, \infty)$, the solution yielded by open-loop information is such that firms produce systematically more than in the static game, as $q^{OL} > q^{CN}$.

On the other hand, the limits appearing in (5.11) reveal the emergence of the perfectly competitive equilibrium if either firms infinitely discount future magnitudes or market price is infinitely sticky (thereby preventing firms from reaching the notional demand at all times).

Summing up, specific limits taken upon the equilibrium magnitudes (typically, but not exclusively, controls) show that the open-loop equilibrium reproduces the equilibrium of the corresponding static game, this being due to the complete absence of state-control loops in the solution method, and independently of the exact formulation of such loops. Yet, this *does not* imply that the open-loop solution can be deemed as static or quasi-static, due to the presence of at least one state equation which, by definition and construction, cannot appear in the static game, and automatically raises the question concerning the trajectory to and the stability of the open-loop solution, the latter being a relevant issue also in the limit, where the static Nash equilibrium pops up.

Additionally, in this setup it clearly emerges that the open-loop solution is only weakly time consistent, and why. The reason is that the open-loop strategy is a function of the state variable but—as we are about to see—cannot replicate any of the infinitely many proper feedback ones, only two of them being linear. The most important implication of this finding, including the properties of the limits appearing in (5.10–5.11), is that *the open-loop strategy almost never coincides with the static Nash strategy*, the latter holding only in the limit, but, in general, not at any time during the game.

### 5.3.2 The Feedback Game

Firm $i$'s HJB equation is

$$\rho V_i(p) = \max_{q_i} \left\{ \left( p - c - \frac{q_i}{2} \right) q_i + V_i'(p) s (\widehat{p} - p) \right\} \tag{5.12}$$

Here, $V_i(p)$ is the value function and $V_i'(p) = \partial V_i(p)/\partial p$ is its partial derivative w.r.t. the price. Solving the first order condition $p - c - q_i - sV_i'(p) = 0$ delivers the optimal output $q^F(p) = p - c - sV_i'(p) > 0$ provided $p > c + sV_i'(p)$. If the latter condition is satisfied, one may impose symmetry on outputs and substitute $q^F(p)$ into (5.12). Then, defining $V(p) = \varepsilon_1 p^2 + \varepsilon_2 p + \varepsilon_3$ implies $V'(p) = 2\varepsilon_1 p + \varepsilon_2$, and solving the related system of Riccati equations, one obtains

$$\varepsilon_3 = \frac{c(c + 4\varepsilon_2 s) + \varepsilon_2 s(2a + 3\varepsilon_2 s)}{2\rho} \tag{5.13}$$

$$\varepsilon_2 = \frac{2\varepsilon_1 s(a + 2c) - c}{\rho + 3s(1 - 2\varepsilon_1 s)} \tag{5.14}$$

$$\varepsilon_1^{\pm} = \frac{6s + \rho \pm \sqrt{(6s + \rho)^2 - 12s^2}}{12s} \tag{5.15}$$

with $\varepsilon_1^{\pm} \in \mathbb{R}$ for all $s, \rho > 0$. The pair $\varepsilon_1^{\pm}$ delivers the pair of linear feedback strategies identified by $q_{\pm}^F = p - c - s\left(2\varepsilon_1^{\pm} p + \varepsilon_2\right)$. A quick look at the phase diagram appearing in Fig. 5.1 suffices to learn that $q_-^F$ is stable, while $q_+^F$ is unstable.

Points $F_-$, $OL$ and $F_+$ identify the three steady states. The graph portrays also the non-invertibility line $q_\infty$ and nonlinear feedback solutions. Concerning the latter, it must be stressed that the open-loop steady-state equilibrium can be

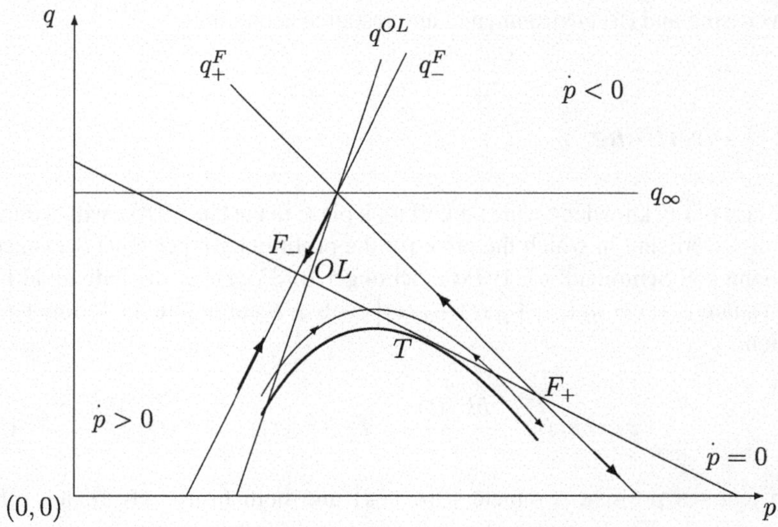


**Fig. 5.1** The phase diagram of the sticky price game

reproduced through a unique nonlinear feedback strategy which never corresponds to $q^{OL}$ except in the steady state. This, by the way, reiterates the conclusion that the open-loop solution of this game is not subgame perfect. In apparently different but formally equivalent terms, one may say that the optimal costate $\lambda$, characterising the optimal open-loop solution at every instant, never coincides with the partial derivative of the value function, $V'$, across the continuum of linear and nonlinear strategies solving the feedback game.

This is very often the case, and can be easily verified in most of LQ games (see Reynolds 1987, 1991; Kobayashi 2015, inter alia), but does not hold in general. And indeed, spotting a game where the optimal costate variable coincides with the partial derivative of the value function (and therefore properly qualifies itself as a *shadow price*) is a precious finding, as it indicates that the game at hand is in fact a perfect game.

Accordingly, the final step consists in taking a close look at this class of games, in some of which the open-loop strategy may indeed be independent of state(s) and therefore may permanently replicate the static Nash equilibrium one, to fully appreciate (i) the inherent advantages of degenerate feedback solutions and, perhaps more importantly, the drawbacks of static analyses "blackboxing" dynamic processes which remain behind the curtains.

## 5.4 Perfect Games

A sensible approach to the exposition of perfect games in many areas of economics and management consists in taking into consideration the two areas in which their emergence has probably delivered the most important and persistent results, namely, (i) advertising and (ii) environmental and resource economics.

### 5.4.1 *Advertising*

To the best of my knowledge, the kickoff took place in the late 1970s, with a duopoly game of advertising in which the price (or the profit margin per unit) is exogenous (Leitmann and Schmitendorf, 1978; Feichtinger, 1983). Here, the individual firm's market share $\varsigma_i(t) = q_i(t) / \left[q_i(t) +_j (t)\right]$ evolves according to the following state equation:

$$\dot{\varsigma}_i = k_i(t) - \frac{bk_i^2(t)}{2} - ck_j(t)\,\varsigma_i(t) - \delta\varsigma_i(t) \qquad (5.16)$$

which features positive parameters $\{b, c, \delta\}$ and both firms' advertising efforts, $k_i(t)$ and $k_i(t)$. The instantaneous individual profit function is $\pi_i(t) = \beta\varsigma_i(t) - k_i(t)$, with parameter $\beta$ scaling revenues (so that the model does not feature a downward sloping market demand). This version of the game turns out to be state-redundant under open-loop information as it satisfies (5.3), and can be easily extended to the oligopoly case (Dragone et al., 2010; Jørgensen et al., 2010).

Note that the model is not defined in LQ form, because of the shape of the state equation. In general, a structure like this would exclude the fully analytical characterisation of the feedback game based upon the solution of the HJB equation, including of course the continuum of nonlinear strategies. Yet, the open-loop solution being subgame perfect and stable, the game delivers a reliable interpretation of what one may expect to see along the saddle path and at the steady-state equilibrium, including the picture of the static Nash equilibrium in the limit.

Moreover, this setup is also extraordinarily versatile. Changing the interpretation of variables to adapt it to other contexts, equally relevant, it may describe an alternative version of Reynolds' (1987; 1991) game of capacity accumulation, where the public authority regulates the price level, or electoral competition for a public office among parties' candidates, in which $\beta$ measures the candidate's appraisal of electoral consensus (Lambertini, 2014). Additionally, the model can be quickly turned into a proper linear state LQ game by rewriting the payoff as $\pi_i(t) = \beta \varsigma_i(t) - k_i^2(t)$ and the state equation as

$$\dot{\varsigma}_i = k_i(t) - c k_j(t) - \delta \varsigma_i(t) \tag{5.17}$$

thereby also preserving most of the qualitative properties of the resulting solution, as well as its intuitive interpretation.

An alternative approach to advertising games granting strong time consistency under open-loop rules dates back to a modified version of the Lanchester-Case model (Lanchester, 1956; Case, 1979) appearing in Sethi and Thompson (1981), Sethi (1983) and Sorger (1989). Once again, the mark-up is constant over the whole horizon, and firm $i$'s market share follows

$$\dot{\varsigma} = k_i(t)\sqrt{1 - \varsigma(t)} - k_j(t)\sqrt{\varsigma(t)} \tag{5.18}$$

while the instantaneous profit function is $\pi_i(t) = \beta \varsigma(t) - b k_i^2(t)$. It turns out that the HJB equation

$$\begin{aligned} &\rho V_i(\varsigma(t)) \\ &= \max_{k_i(t)} \left\{ \beta \varsigma(t) - b k_i^2(t) + V_i'(\varsigma(t)) \left[ k_i(t)\sqrt{1 - \varsigma(t)} - k_j(t)\sqrt{\varsigma(t)} \right] \right\} \end{aligned} \tag{5.19}$$

can be solved to reproduce the open-loop equilibrium of the corresponding Hamiltonian by adopting a value function specified as $V_i(\sigma(t)) = \varepsilon_{i1}\sigma(t) + \varepsilon_{i2}$, although the game is neither state-linear nor linear-quadratic.

Another approach exhibiting the same property is an oligopoly game of goodwill in the vein of Fershtman (1984), among many others, appearing in Lambertini and

Zaccour (2015). The demand structure is borrowed from Singh and Vives (1984), with $n$ symmetric single-product firms in prices or quantities. Under Cournot behaviour, individual market demands are $p_i(t) = a - q_i(t) - \gamma Q_{-i}(t)$, $Q_{-i}(t) = \sum_{j \neq i} q_j(t)$ being the aggregate output of the $n-1$ rivals, while parameter $\gamma \in [0, 1]$ measures the degree of substitutability. The individual profit function is

$$\pi_i(t) = G_i(t) q_i(t) - bk_i^2(t) \tag{5.20}$$

in which $R_i(t) = p_i(t) q_i(t)$ is the revenue and $G_i(t)$ is the brand equity of variety $i$, which evolves according to

$$\dot{G}_i \equiv \frac{dG_i(t)}{dt} = k_i(t) - \sigma K_{-i}(t) - \delta G_i(t) \tag{5.21}$$

Parameter $\sigma \in [0, 1/(n-1)]$ scales the intensity of the pressure posed by the rivals' collective advertising effort. However, the instantaneous payoff is the product of the state times a quadratic function of the outputs or prices, depending on the nature of market competition. Since the market game can be solved quasi-statically and the Hamiltonian is additively separable w.r.t. states and advertising controls, the open-loop solution is also a degenerate feedback one and can be replicated solving the HJB equation.

The latter model and its interpretation offer an ideal ground to test and appreciate the relevance of subgame perfect open-loop solution. The static version of the game is indeed a toy version of the dynamic one (the same is not true for the other advertising games discussed above) and lends itself to be used as a test-bed of many of the considerations formulated across this chapter, in particular the relevance of manageable open-loop analyses being also illuminating and convincing, well beyond the fact that in the limit they deliver the static outcome. Confining ourselves to this specific detail would amount to disregarding the full-fledged range of informations contained in the dynamics of the game along the path to the long-run equilibrium. And we would also miss the ultimate implication of this discussion, which neatly emerges from the analysis carried out in the next section.

### 5.4.2 Environment and Natural Resources

Let's consider a differential game unravelling over $t \in [0, \infty)$, in which $n \geq 1$ firms play *à la* Cournot along the demand function $p(t) = a - \sum_{i=1}^{n} q_i(t)$ using a common cost function $C_i(t) = cq_i(t) + bq_i^2(t)$, and facing a single state equation

$$\dot{X} = \alpha \sum_{i=1}^{n} q_i(t) - \delta X(t) \tag{5.22}$$

If parameters $\alpha$ and $\delta$ are both positive, the model captures the environmental consequences of firms' strategic behaviour, and (5.22) describes the motion of the stock of GHGs, $X(t)$, net of the absorption rate of the planet's natural carbon sinks, $\delta$. In the opposite case, if $\alpha$ and $\delta$ are both negative, (5.22) is a linear approximation of the pseudo-exponential growth of a living stock being extracted to become a final good on consumer markets (as in Fujiwara 2008; and Lambertini and Mantovani (2014, 2016), *inter alia*). In the latter case, the proper formulation of the stock's dynamics should include the downward sloping part of the full piecewise approximation of the instantaneous growth rate, as in Benchekroun (2003, 2008) and Colombo and Labrecciosa (2015) and several others. For the moment, we may leave this aspect aside, as it will be taken explicitly into consideration below.

The individual firm's profit function can be defined as

$$\pi_i(t) = [p(t) - c - \tau - bq_i(t)]\, q_i(t) \tag{5.23}$$

in which $\tau > 0$ is a tax (on GHG emissions) or a tariff (on extraction), and the lack of indication of the time argument means that this policy instrument can be seen as a constant, increasing marginal costs by the same amount at all times. This is functional to the aim of the present discussion, but of course the extant literature has endogenised it (see Benchekroun and Long 1998).[4] More on this aspect below.

Without delving explicitly into analytical details, we may grasp the essential elements of the nature of the game by noting the formal properties of the HJB, and those of the equilibrium outcomes by looking at the resulting phase diagram.

To begin with, the relevant HJB equation is

$$\rho V_i(X) = \max_{q_i} \left\{ [p(t) - c - \tau - bq_i(t)]\, q_i(t) + V_i'(X) \left[ \alpha \sum_{i=1}^{n} q_i(t) - \delta X(t) \right] \right\} \tag{5.24}$$

and the expression appearing on the r.h.s. is evidently linear in the state variable. Moreover, the instantaneous profit function is defined in the space of controls only. Taken together, these features of (5.24) immediately imply that (i) the open-loop strategy is independent of the state variable and (ii) it is indeed a degenerate feedback one.

Solving the game to identify the continuum of feedback strategies (see Lambertini, 2018, pp. 189–94), one obtains the phase diagram appearing in Fig. 5.2. The dynamic properties have been intentionally omitted, except for the indication of the sign of $\dot{X}$, which determines the portion of stable equilibria along the segment $(A, B)$: if $\alpha, \delta > 0$, any steady state in $(T, B]$ (resp., $[A, T)$) is stable (unstable),

[4] A reconstruction of the wide stream of research in environmental and natural resource economics based on dynamic models using these structures can be found in Long (2011) and Lambertini (2013, 2018).

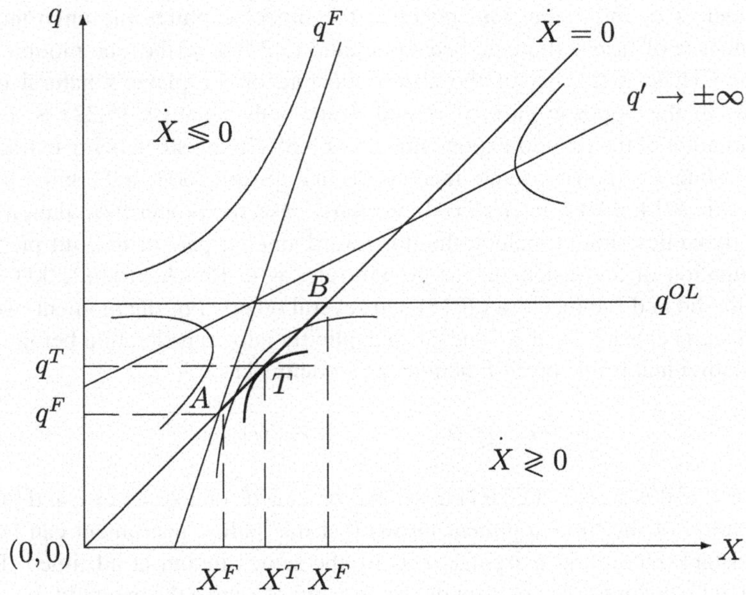


**Fig. 5.2** The phase diagram of the dual-purpose game

while the opposite holds for all $\alpha, \delta > 0$. In both cases, the tangency point at $T$ is semi-stable.

Accordingly, the 'flat' open-loop strategy is stable and strongly time consistent if the game deals with GHG emissions, while being unstable when the matter is the economic exploitation of a living stock. This fact tells us another relevant piece of information, which can be formulated as follows: in itself, finding out that the open-loop strategy coincides with the static one at all times does not suffice to conclude that the static game may reproduce the essence of its dynamic version, for two equally relevant reasons. The first is that the static solution may at most identify one of the infinitely many feedback equilibrium strategies; the second is that it may reproduce an unstable feedback strategy, as is the case if the model discusses natural resource extraction. Be that as it may—that is, even when the open-loop solution is stable—it remains true that a static game approach in an inherently dynamic problem is literally incapable of collecting an arbitrarily large amount of information concerning the possible equilibrium configurations of the dynamic system being blackboxed, the property thereof, and the detailed description of the path to any of its infinitely many equilibria.

What if the tax/tariff is endogenised? This poses another intriguing problem concerning the time consistency of any such policy. This issue is brilliantly solved by Benchekroun and Long (1998) specifying the emission tax as a linear function of the state variable (GHG emissions). Yet, this additional feature causes the open-loop solution to be no longer Markov-perfect, as the resulting objective functional is not additively separable in state and controls, which, by the way, means that in general the welfare-maximising tax engendered in static oligopoly games does not coincide with the strongly time consistent tax identified by Benchekroun and Long (1998). Another route, preserving the subgame perfectness of open-loop rules, is taken in Feichtinger et al. (2016), stipulating that the tax (i) is accompanied by a regulated price and (ii) is used to maximise welfare at the steady state. This approach is used to include R&D for abatement technologies and to show that the aggregate R&D effort is concave and single-peaked in the number of firms. This is an example—by no means the only one—of models in which the presence of an appropriate regulatory toolkit transforms the game into a perfect one, while its original formulation wouldn't be so.

The last step brings us to Ramsey-Lotka-Volterra games (Lotka, 1925; Ramsey, 1928; Volterra, 1931), which include both capacity accumulation games (Cellini and Lambertini, 1998, 2008) and resource extraction games (Lambertini and Leitmann, 2019; Feichtinger et al., 2022). Irrespective of the specific nature of the game, this model features a properly concave and single-peaked growth rate of the stock, continuously twice differentiable, instead of a linear or piecewise linear approximation of it, with a kink.

Since we are treating renewables, we may suppose firms exploit a living stock, say, *à la* Cournot, whereby the relevant state equation is

$$\dot{X} = \delta X(t)[1 - \beta X(t)] - \gamma Q(t) \tag{5.25}$$

in which parameter $\beta$ is the inverse of the habitat's carrying capacity. Demand is the same as above, and the profit function is $\pi_i(t) = p(t) q_i(t) - C_i(t)$. Now, it is pretty obvious that the Hamiltonian function

$$\mathcal{H}_i(X(t)) = e^{-\rho t}\{p(t) q_i(t) - C_i(t) + \lambda_i(t)[\delta X(t)(1 - \beta X(t)) - \gamma Q(t)]\} \tag{5.26}$$

does not feature a linear state game. Indeed, the solution of the game reached on the basis of Pontryagin's maximum principle delivers the picture in Fig. 5.3, where the vertical line corresponding to the Ramsey rule falls systematically short of the maximum sustainable yield, i.e., the peak of the resource's growth rate. The Cournot

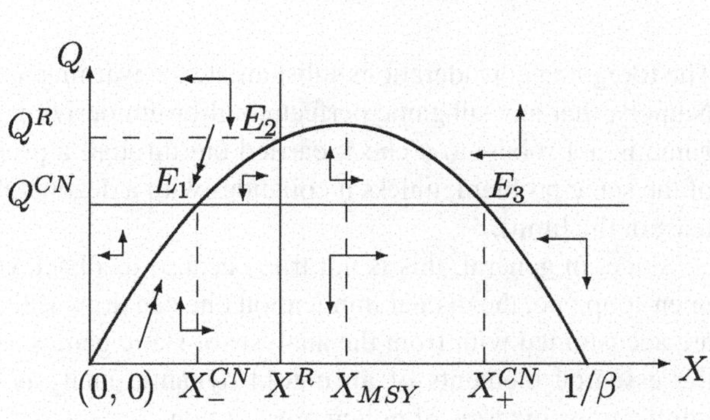


**Fig. 5.3** The full phase diagram including the Ramsey rule

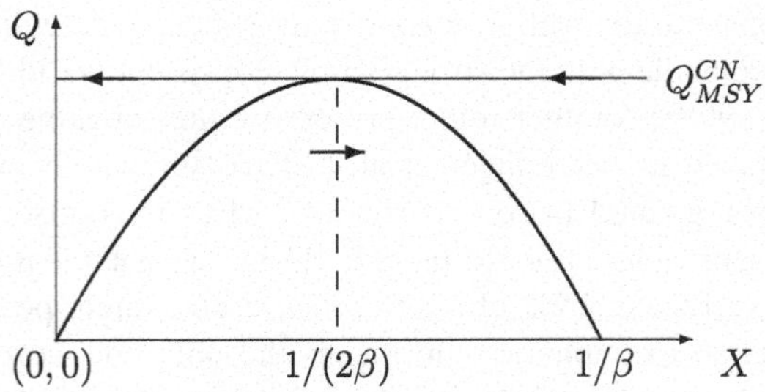


**Fig. 5.4** The degenerate feedback harvest with regulated access

harvest delivers the saddle point equilibrium at $E_1$ if, as in the situation appearing in Fig. 5.3, it is lower than the Ramsey harvest at point $E_2$ (unstable, while $E_3$ is stable).

However, as we know since Leitmann (1973) and Goh et al. (1974), the costate equation of this game is differential equation in separable variables admitting the solution $\lambda_i(t) = 0$ at all times for all players, as can be easily verified on the basis of (5.26). This implies that the Ramsey rule fades away completely, making room for a regulated access to the common pool which consists in limiting entry in correspondence of the number of firms driving the industry harvest as close as possible to the maximum sustainable yield. The number being perforce an integer, the public authority may use complementary instruments (interest rates, taxes, subsidies) to achieve its goal. This solution delivers the phase diagram in Fig. 5.4, where harvesting at the maximum sustainable yield happens in correspondence of the semi-stable tangency point.

This setup illustrates that a whole class of games governed by nonlinear state equations may reveal its state-redundant nature and yield a reliable solution of a relevant and long-lasting problem—as the preservation of biological stocks—through the fruitful exploitation of firms' myopic behaviour. All of this holds in spite of the fact that the setup, by construction, does not facilitate the formulation of an appropriate guess about the shape of the value function.

## 5.5 Concluding Remarks

The foregoing considerations substantiate the warning appearing in the introduction. Namely, that any subgame perfect equilibrium deriving from a possibly multistage game is not robust to a check carried out through a properly dynamic formulation of the same problem, unless it coincides with a degenerate feedback equilibrium (at least in the limit).

Since, in general, this is not true because feedback equilibria rarely include the open-loop one, the dismal implication one can draw is that most of the equilibria we are accustomed with from the analysis of static games are not snapshots containing the essential elements of an untold dynamic analysis, with obvious undesirable implications in terms of policy prescription.

Hence, the search for relevant and insightful games delivering strongly time consistent solutions under open-loop information should remain at the top of the agenda and should also inform any further developments of applied research relying on static games in the whole range of social sciences, the latter becoming permanently aware of the need of looking for confirmations from the 'dynamic side'.

## References


Benchekroun, H. (2003). Unilateral production restrictions in a dynamic duopoly. *Journal of Economic Theory, 111*, 214–239.

Benchekroun, H. (2008). Comparative dynamics in a productive asset oligopoly. *Journal of Economic Theory, 138*, 237–261.

Benchekroun, H., & Long, N. V. (1998). Efficiency inducing taxation for polluting oligopolists. *Journal of Public Economics, 70*, 325–342.

Case, J. (1979). *Economics and the competitive process*. New York University Press.

Cellini, R., & Lambertini, L. (1998). A dynamic model of differentiated oligopoly with capital accumulation. *Journal of Economic Theory, 83*, 145–155.

Cellini, R., & Lambertini, L. (2004). Dynamic oligopoly with sticky prices: Closed-loop, feedback and open-loop solutions. *Journal of Dynamical and Control Systems, 10*, 303–314.

Cellini, R., & Lambertini, L. (2007). A differential oligopoly game with differentiated goods and sticky prices. *European Journal of Operational Research, 176*, 1131–1144.

Cellini, R., & Lambertini, L. (2008). Weak and strong time consistency in a differential oligopoly game with capital accumulation. *Journal of Optimization Theory and Applications, 138*, 17–26.

Cellini, R., Lambertini, L., & Leitmann, G. (2005). Degenerate feedback and time consistency in differential games. In E. P. Hofer & E. Reithmeier (Eds.), *Modeling and Control of Autonomous Decision Support Based Systems. Proceedings of the 13th International Workshop on Dynamics and Control* (pp. 185–192). Shaker Verlag.

Clemhout, S., & Wan, Jr. H. Y. (1974). A class of trilinear differential games. *Journal of Optimization Theory and Applications, 14*, 419–424.

Clemhout, S., & Wan, Jr. H. Y. (1979). Interactive economic dynamics and differential games. *Journal of Optimization Theory and Applications, 27*, 7–30.

Colombo, L., & Labrecciosa, P. (2015). On the Markovian efficiency of bertrand and cournot equilibria. *Journal of Economic Theory, 155*, 332–358.

Dockner, E. J, Feichtinger, G., & Jørgensen, S. (1985). Tractable classes of nonzero-sum open-loop nash differential games: Theory and examples. *Journal of Optimization Theory and Applications, 45*, 179–197.

Dockner, E. J, Jørgensen, S., Van Long, N., & Sorger, G. (2000). *Differential games in economics and management science*. Cambridge University Press.

Dragone, D., Lambertini, L., & Palestini, A. (2010). The Leitmann-Schmitendorf advertising game with n players and time discounting. *Applied Mathematics and Computation, 217*, 1010–1016.

Feichtinger, G. (1983). The nash solution of an advertising differential game: Generalization of a model by Leitmann and Schmitendorf. *IEEE Transactions on Automatic Control, 28*, 1044–1048.

Feichtinger, G., Lambertini, L., Leitmann, G., & Wrzaczek, S. (2016). R&D for green technologies in a dynamic oligopoly: Schumpeter, arrow and inverted U's , *European Journal of Operational Research, 249*, 1131–1138.

Feichtinger, G., Lambertini, L., Leitmann, G., & Wrzaczek, S. (2022). Managing the tragedy of commons and polluting emissions: A unified view. *European Journal of Operational Research, 303*, 487–499.

Fershtman, C. (1984). Goodwill and market shares in oligopoly. *Economica, 51*, 271–281.

Fershtman, C. (1987). Identification of classes of differential games for which the open-loop is a degenerate feedback nash equilibrium. *Journal of Optimization Theory and Applications, 55*, 217–231.
Fershtman, C., & Kamien, M. (1987). Dynamic duopolistic competition with sticky prices. *Econometrica, 55*, 1151–1164.
Fujiwara, K. (2008). Duopoly can be more anti-competitive than monopoly. *Economics Letters, 101*, 217–219.
Goh, B., Leitmann, G., & Vincent, T. (1974). Optimal control of the prey-predator system. *Journal of Mathematical Biosciences, 19*, 263–286.
Jørgensen, S., Martín-Herrán, G., & Zaccour, G. (2010). The Leitmann-Schmitendorf advertising differential game. *Applied Mathematics and Computation, 217*, 1110–1116.
Kobayashi, S. (2015). On a dynamic model of cooperative and noncooperative R and D in oligopoly with spillovers. *Dynamic Games and Applications, 5*, 599–619.
Lambertini, L. (2013). *Oligopoly, the environment and natural resources*. Routledge.
Lambertini, L. (2014). Dynamic analysis of an electoral campaign. In J. Haunschmied, V. Veliov & S. Wrzaczek (Eds.), *Dynamic games in economics* (pp. 187–204). Springer.
Lambertini, L. (2018). *Differential games in industrial economics*, Cambridge University Press.
Lambertini, L., & Leitmann, G. (2019). On the attainment of the maximum sustainable yield in the Verhulst-Lotka-Volterra model. *Automatica, 110*, Article 108555, 1–5.
Lambertini, L., & Mantovani, A. (2014). Feedback equilibria in a dynamic renewable resource oligopoly: Pre-emption, voracity and exhaustion. *Journal of Economic Dynamics and Control, 47*, 115–122.
Lambertini, L., & Mantovani, A. (2016). On the (in)stability of nonlinear feedback solutions in a dynamic duopoly with renewable resource exploitation. *Economics Letters, 143*, 9–12.
Lambertini, L., & Zaccour, G. (2015). Inverted-U aggregate investment curves in a dynamic game of advertising. *Economics Letters, 132*, 34–38.
Lanchester F. W. (1956). Mathematics in warfare. In J. R. Newman (Ed.), *The world of mathematics*. Simon and Schuster.
Leitmann, G. (1973). A minimum principle for a population equation. *Journal of Optimization Theory and Applications, 9*, 155–156.
Leitmann, G., & Schmitendorf, W. E. (1978). Profit maximization through advertising: A nonzero sum differential game approach. *IEEE Transactions on Automatic Control, 23*, 646–650.
Long, N. V. (2011). Dynamic games in the economics of natural resources: A survey. *Dynamic Games and Applications, 1*, 115–148.
Lotka, A. J. (1925). *Elements of physical biology*. Williams and Wilkins.
Mehlmann, A. (1988). *Applied differential games*. Plenum Press.
Mehlmann, A., & Willing, R. (1983). On nonunique closed-loop nash equilibria for a class of differential games with a unique and degenerate feedback solution. *Journal of Optimization Theory and Applications, 41*, 463–472.
Ramsey, F. P. (1928). A mathematical theory of saving. *Economic Journal, 38*, 543–559.
Reinganum, J. (1982a). A class of differential games for which the closed loop and open loop nash equilibria coincide. *Journal of Optimization Theory and Applications, 36*, 253–262.
Reinganum, J. (1982b). A dynamic game of R&D: Patent protection and competitive behavior. *Econometrica, 50*, 671–688.
Reynolds, S.: (1987). Capacity investment, preemption and commitment in an infinite horizon model. *International Economic Review, 28*, 69–88.
Reynolds, S. (1991). Dynamic oligopoly with capacity adjustment costs. *Journal of Economic Dynamics and Control, 15*, 491–514.
Sethi, S. (1983). Deterministic and stochastic optimization of a dynamic advertising model. *Optimal Control Applications and Methods, 4*, 179–184.
Sethi, S., & Thompson, G. (1981). *Optimal control theory: Applications to management science*. Nijhoff.
Simaan, M., & Takayama, T. (1976). Optimum monopolist control in a dynamic market. *IEEE Transactions on Systems, Man, and Cybernetics, SMC-6*, 799–807.

Simaan, M., & Takayama, T. (1978). Game theory applied to dynamic duopoly problems with production constraints. *Automatica, 14*, 161–166.

Singh, N., & Vives, X. (1984). Price and quantity competition in a differentiated duopoly. *RAND Journal of Economics, 15*, 546–554.

Sorger, G. (1989). Competitive dynamic advertising: A modification of the case game. *Journal of Economic Dynamics and Control, 13*, 55–80.

Tsutsui, S., & Mino, K. (1990). Nonlinear strategies in dynamic duopolistic competition with sticky prices. *Journal of Economic Theory, 52*, 136–161.

Volterra, V. (1931). Variations and fluctuations of the number of individuals in animal speciesliving together. In R. N. Chapman (Ed.), *Animal ecology* (pp. 31–113). McGraw-Hill.

# Chapter 6
# A Linear State Game of Advertising *à la* Vidale-Wolfe

**Luca Lambertini and Andrea Mantovani**

**Abstract** We revisit the tradition of advertising models stemming from Vidale and Wolfe (Oper Res 5:370–381;1957) to illustrate the possibility of building up a game delivering a (degenerate) feedback equilibrium under open-loop rules. To this aim, we reformulate the state equation of the generic firm in such a way that its own advertising effort and the rivals' reaction to it enter the state dynamics additively. This approach amounts to envisaging situations where advertising has an essentially predatory/defensive nature, as it is not designed to modify the natural growth rate of a firm's sales or market share. This modeling strategy gives the game a state-linear structure, which also delivers an Arrovian result concerning the relationship between the aggregate advertising effort and industry structure.

**Keywords** Advertising · Oligopoly · Differential games · Strong time consistency

**JEL Codes** C73, L13, M37

## 6.1 Introduction

Given its inherently dynamic nature, advertising stands out as one of the most debated topics within the realm of optimal control and differential game theory since seminal works of Friedman (1958), Clemhout et al. (1971), and Leitmann and Schmitendorf (1978), among others. Notably, advertising efforts are typically

---

L. Lambertini
Department of Economics and Alma Climate Centre, University of Bologna, Bologna, Italy
e-mail: luca.lambertini@unibo.it

A. Mantovani (✉)
Department of Information, Operations and Management Sciences, TBS Business School, Toulouse, France
e-mail: a.mantovani@tbs-education.fr



F. El Ouardighi, G. Feichtinger (eds.), *The Unaffordable Price of Static Decision-making Models*, International Series in Operations Research & Management Science 365, https://doi.org/10.1007/978-3-031-88638-6_6

categorized into three main types: informative, persuasive, and complementary, as outlined by Stigler and Becker (1977) and Becker and Murphy (1993). The latter category suggests that advertising plays a role in defining the overall features of a product, thus complementing it. In the applications of dynamic techniques, the focus has often been directed toward the specific state variable affected by advertising efforts or on examining the impact of advertising throughout the product life cycle.

Dynamic models addressing demand (or output) expansion are often associated with the concept of persuasive advertising. These models typically involve firms investing to increase choke prices, which become the relevant state variable, as in Cellini and Lambertini (2003a,b) and Cellini et al. (2008). Alternatively, other models consider output (or sales) levels as the relevant state variables reacting to firms' advertising efforts, as exemplified by Vidale and Wolfe (1957) and its many follow-ups.

Another scenario is that in which advertising aimed at enhancing goodwill, as in Nerlove and Arrow (1962), Gould (1970), Fershtman (1984), and many others, where the individual firm's profit is augmented by a state variable inflating revenues or gross profits. Additionally, in market growth or product diffusion models, such as those pioneered by Bass (1969), the relevant state variable is a firm's cumulative volume of sales. Notably, models belonging to this latter class are often characterized by a finite time horizon, as further product diffusion is inevitably constrained by the availability of competing goods from existing rivals or new entrants.

The wide literature concerning the dynamic analysis of advertising in competitive markets features several examples of games generating open-loop solutions which are Markov-perfect. This is the case, for instance, in Leitmann and Schmitendorf (1978), Feichtinger (1983), and other contributions sharing the property of state redundancy although not being state-linear, including some formulations of games with advertising for goodwill, as in Lambertini and Zaccour (2015).[1]

In this paper, we revisit the tradition of advertising models originating from Vidale and Wolfe (1957) to show that it is possible to build a game that achieves a (degenerate) feedback equilibrium under open-loop rules. In line with Vidale and Wolfe (1957), previous formalizations of the cross-effects of firms' advertising efforts and sales across the set of state variables were characterized by the lack of additive separability between states and controls, as in the competitive version of the model appearing in Deal (1979). Consequently, a fully analytical characterization of feedback equilibria through the solution of the relevant set of Hamilton-Jacobi-Bellman equations remains out of reach as one cannot formulate a plausible guess about the shape of the relevant value function. This limitation also yields an open-loop solution which is weakly time consistent.

[1] It must also be stressed that there exists a class of advertising games based upon Lanchester (1956) and Case (1979) whose formulation is neither state-linear nor linear-quadratic, which nonetheless can be analytically solved in feedback strategies using linear value functions. See Sethi and Thompson (1981), Sethi (1983) and Sorger (1989), *inter alia*.

We reformulate the state equation of the generic firm in such a way that its own advertising effort and the rivals' best replies to it enter the state dynamics additively. We preserve the role of the carrying capacity while obtaining an open-loop solution which is indeed a degenerate feedback one. This approach enables us to characterize situations where advertising has an essentially dual role, being both predatory and defensive at the same time, as it is not designed to directly modify the natural growth rate of a firm's sales or market share. This modeling strategy gives the game a state-linear structure, which also yields an Arrovian result concerning the relationship between the aggregate advertising effort and industry structure.

There remain to add a few relevant elements about a crucial aspect of the model and the nature of its equilibrium outcome. The results attained through the solution of this advertising game have some interesting implications concerning the potential, too often overlooked, of the open-loop equilibrium concept and the very fact that open-loop information be assumed, the common objection being that doing so leads to the replication of static equilibria. Indeed, this is not the case in general, a fortiori in games sharing the property we are referring here. More explicitly, the ensuing analysis hinges upon the possibility of specifying the setup (in this case, belonging to a well-established tradition which is relevant for industrial economics and business and management alike) so as to capture a plausible and relevant real-world scenario and enjoying the property of state linearity, yielding strong time consistency under open-loop rules. This makes room for a properly dynamic characterization of firms' behavior as well as the aggregate performance of a whole industry, through feedback rules emerging from the Hamiltonian formulation of the game itself, thereby validating the adoption of the open-loop approach.

The remainder of the chapter is structured as follows. Section 6.2 contains the motivation and layout of the game. The Markov-perfect open-loop solution and the analysis of its stability properties is in Sect. 6.3. A few concluding remarks are in Sect. 6.4.

## 6.2 Setup

The differential game describes a sales-response model *à la* Vidale and Wolfe (1957), which has in common with Bass (1969) and the ensuing literature of the presence of state equations mimicking a logistic growth curve, although we look at the case of non-durables.[2] In particular, we propose a modified version of Deal's (1979), in terms of the formalization of the cross-effects of firms' advertising efforts and sales across the set of the state variables.

Consider a population of firms $\mathcal{N} = \{1, 2, 3, \ldots n\}$ endowed with the same technology, summarized by a constant marginal production cost $c > 0$. Firms sell

[2] For overviews of the related literature, see Dockner et al. (2000), Erickson (2003), Jørgensen and Zaccour (2004), and Lambertini (2018).

a homogeneous good over continuous time $t \in [0, \infty)$. The individual volume of instantaneous sales is $x_i(t) \geq 0$, $i = 1, 2, \ldots n$, and the unit margin is $P = p - c > 0$. Here, we posit that either the sector is perfectly competitive (so that firms permanently face an infinitely elastic demand function at $p$) or price is regulated (and time-invariant, as is the case of marginal cost). At every instant, each firm may boost its sales volume through an advertising effort $k_i(t)$, which entails an instantaneous cost $\Gamma_i(t) = bk_i^2(t)$, where $b$ is a positive constant. Therefore, the individual firm's instantaneous profit function is $\pi_i(t) = (p - c)x_i(t) - \Gamma_i(t) = Px_i(t) - bk_i^2(t)$.

Before defining the state dynamics, it is useful to briefly discuss how it has been specified in the extant literature and what implications this has engendered. In Deal (1979), originally formulated as a duopoly model, the individual state equation is

$$\dot{x}_i(t) = \zeta k_i(t)\left[1 - \frac{x_i(t) + X_{-i}(t)}{X_{\max}}\right] - \delta x_i(t), \tag{6.1}$$

which can be extended to include the negative effect exerted by the $n - 1$ rivals, $K_{-i}(t) = \sum_{i \neq j} k_j(t)$:

$$\dot{x}_i(t) = \zeta\left[k_i(t) - \beta K_{-i}(t)\right]\left[1 - \frac{x_i(t) + X_{-i}(t)}{X_{\max}}\right] - \delta x_i(t), \tag{6.2}$$

where $X_{\max} \geq \sum_{i=1}^{n} x_i(t)$ is the maximum volume of output consumers may absorb from this industry[3] and $\{\beta, \delta, \zeta\}$ are positive constants. In particular, $\zeta$ is the "natural" growth rate of firm $i$'s share. In (6.1) and (6.2), states appear at the first degree, and therefore the instantaneous growth rate is not implying a logistic growth, just because the element that would imply it is replaced by a function of advertising controls, either $k_i(t)$ or $k_i(t) - \beta K_{-i}(t)$.

A few additional words are in order, concerning the intensity of the negative effect of the rivals's advertising campaigns. In line with the parallel literature on R&D for process innovation with technological spillovers, stemming from d'Aspremont and Jacquemin (1988), it seems appropriate to assume $\beta \in [0, 1/(n-1)]$, in such a way that the instantaneous impact of $K_{-i}(t)$ is at most as large as that of firm $i$'s own effort. And indeed, after solving the game, we will see that there exists a solid reason—not directly related to firms' interplay in the R&D space—to adopt

**Assumption A** $\beta \in [0, 1/(n-1)]$.

Moreover, we also introduce

[3] Borrowing from the jargon of the parallel literature on biological natural resources, $X_{\max}$ is the industry's *carrying capacity* (see, among many others, Clark 1990).

**Assumption B** $X_{\max} \in \left(0, \overline{X}_{\max}\right)$, with

$$\overline{X}_{\max} \equiv \frac{nP\left[1-\beta\left(n-1\right)\right]-2bn\zeta\left(\delta+\rho\right)+\sqrt{\Omega}}{4b\left(\delta+\rho\right)^{2}} \in \mathbb{R}^{+},$$

$$\Omega \equiv n[8bP\left(1+\beta\right)\left(1-\beta\left(n-1\right)\right)\left(n-1\right)\zeta\left(\delta+\rho\right)+ \\ n(P\left(1-\beta\left(n-1\right)\right)+2b\zeta\left(\delta+\rho\right)^{2}].$$

The role of Assumption B is to ensure $\sum_{i=1}^{n} x_i\left(t\right)/X_{\max} < 1$ throughout the game as well as at the steady-state equilibrium, thereby excluding the arising of a corner solution with demand saturation at all times.

Now we may focus on the difference between (6.1) and (6.2). While in the former the presence of rivals is signaled by their sales levels only, in the latter it also takes the form of the countervailing effect associated to advertising efforts (which may have, e.g., comparative nature). However, both specifications of the state equation(s) have a fundamental implication as far as the solvability of the game under feedback rules is concerned. This is due to the lack of additive separability between states and controls in both (6.1) and (6.2), which are clearly nonlinear and therefore do not allow for either an intuitive conjecture of the value function or, consequently, for a fully analytical characterization of feedback equilibria through the solution of the relevant set of Hamilton-Jacobi-Bellman equations. And, of course, the specification of state equations as in either (6.1) or (6.2) makes the open-loop solution weakly time consistent.

This has triggered several efforts aimed at delivering models with largely although not entirely equivalent economic interpretations, but producing strongly time consistent equilibria, possibly in the form of degenerate feedback strategies designed under open-loop information (see, e.g., Leitmann and Schmitendorf 1978; Feichtinger 1983; Dragone et al. 2010; and Jørgensen et al. 2010), by formulating state equations in such a way that the game becomes state-redundant, at least in its open-loop form.

Yet, another avenue—which, to the best of our knowledge, has been overlooked thus far—is open for interesting extensions. This consists in constructing additively separable state equations in which the role of carrying capacity is preserved and, nonetheless, the open-loop solution is indeed a degenerate feedback one. To this purpose, we specify the state dynamics as follows:

$$\dot{x}_i\left(t\right)=\zeta\left[1-\frac{x_i\left(t\right)+X_{-i}\left(t\right)}{X_{\max}}\right]+k_i\left(t\right)-\beta K_{-i}\left(t\right)-\delta x_i\left(t\right). \tag{6.3}$$

Here, the whole net advertising effort adds up to the "natural" growth rate of the individual firm's sales, in the same way as the harvest rate of firms in a renewable resource extraction game (Lambertini and Leitmann, 2019). Since the control used in this game is not a harvest rate, the interpretation of (6.3) is the following. The $n-1$

rivals are aware that their advertising efforts may exert a business stealing effect by shifting downward firm $i$ sales, all else equal (specifically, for any given $\delta$ and $\zeta$), and therefore firm $i$ reacts to diminish the impact of the negative spillover associated with $K_{-i}(t)$. Naturally, firm $i$ also knows that, by doing so, it is impacting each of the rivals' sales in an analogous way. In a sense, this formulation has some features in common with the concept of *conformance quality* dating back to Garvin (1988), which appears in an additively separable way in analogous extensions of the Vidale-Wolfe model (see, e.g., El Ouardighi et al. 2013).

Accordingly, the Hamiltonian function of firm $i$ is

$$\mathcal{H}_i\left(x_i(t), X_{-i}(t), k_i(t), K_{-i}(t), \lambda_{ij}(t)\right) = e^{-\rho t}\left\{Px_i(t) - bk_i^2(t) + \right.$$
$$\lambda_{ii}(t)\left[\zeta\left(1 - \frac{x_i(t) + X_{-i}(t)}{X_{\max}}\right) + k_i(t) - \beta K_{-i}(t) - \delta x_i(t)\right] +$$
$$\left.\sum_{j\neq i}\lambda_{ij}(t)\left[\zeta\left(1 - \frac{x_j(t) + X_{-j}(t)}{X_{\max}}\right) + k_j(t) - \beta K_{-j}(t) - \delta x_j(t)\right]\right\}, \tag{6.4}$$

where $\lambda_{ij}(t)$ is the relevant capitalized costate variable, for all $i$ and $j$. We are now ready to illustrate the solution of the game under open-loop information and its key properties.

## 6.3 Solving the Game

The individual firm's first-order condition (FOC) w.r.t. $k_i(t)$ is

$$\lambda_{ii}(t) - 2bk_i(t) - \beta\sum_{j\neq i}\lambda_{ij}(t) = 0, \tag{6.5}$$

which is accompanied by the following set of costate equations:

$$-\frac{\partial\mathcal{H}_i(\cdot)}{\partial x_i(t)} = \dot{\lambda}_{ii}(t) - \rho\lambda_{ii}(t) \Rightarrow$$
$$\dot{\lambda}_{ii}(t) = \frac{\left[\lambda_i(t)(\delta+\rho) - P\right]X_{\max} + \zeta\left[\lambda_{ii}(t) + \sum_{j\neq i}\lambda_{ij}(t)\right]}{X_{\max}} \tag{6.6}$$

$$-\frac{\partial\mathcal{H}_i(\cdot)}{\partial x_j(t)} = \dot{\lambda}_{ij}(t) - \rho\lambda_{ij}(t) \Rightarrow$$
$$\dot{\lambda}_{ij}(t) = \frac{\lambda_{ij}(t)(\delta+\rho)X_{\max} + \zeta\left[\lambda_{ii}(t) + \sum_{j\neq i}\lambda_{ij}(t)\right]}{X_{\max}}, \tag{6.7}$$

while the set of transversality conditions is summarized by

$$\lim_{t\to\infty} e^{-\rho t}\lambda_{ij}(t)\, x_j(t) = 0\,\forall\, j = 1, 2, 3, \ldots n. \tag{6.8}$$

We are now in a position to assess some essential properties of the open-loop solution by looking at the system (6.5–6.7). To begin with, (6.5) does not explicitly feature any state variable, and this is true also for all costate equations (6.6–6.7). Intuitively, this is due to the fact that the present game is linear in states, thanks to its additively separable reformulation. These elements prove the following:

**Lemma 6.1** *Since the game has a state-linear structure, its open-loop solution is a degenerate feedback one.*

There remains to analytically characterize it. To this aim, we may (i) drop the time argument for the sake of brevity and (ii) impose symmetry upon all variables not pertaining to firm $i$ and solve (6.5) to find the expression of the optimal $\lambda_{ii}(t)$ at a generic instant,

$$\lambda_{ii} = 2bk_i + \beta(n-1)\lambda_{ij}, \tag{6.9}$$

and then solve the same equation again w.r.t. $k_i$, to get

$$k_i = \frac{\lambda_{ii} - \beta(n-1)\lambda_{ij}}{2b}. \tag{6.10}$$

The above expression can be differentiated w.r.t. time to deliver the following control equation:

$$\dot{k}_i = \frac{\dot{\lambda}_{ii} - \beta(n-1)\dot{\lambda}_{ij}}{2b}, \tag{6.11}$$

which, using (6.6–6.9), simplifies as follows:

$$\dot{k}_i = \frac{(2bk_i - P)X_{\max} - \xi\left[1-\beta(n-1)\right]\left[2bk_i + (1+\beta)(n-1)\right]\lambda_{ij}}{2bX_{\max}}. \tag{6.12}$$

Then, we may solve the system (6.7-6.12) to find the expressions of the optimal $k_i$ and $\lambda_{ij}$ at any point in time. This is done by posing equal to zero the related integration constants, thus obtaining

$$k_i^* = \frac{P\left[(\delta+\rho)X_{\max} + \zeta(1+\beta)(n-1)\right]}{2b(\delta+\rho)\left[(\delta+\rho)X_{\max} + n\zeta\right]}$$

$$\lambda_{ij}^* = -\frac{P\zeta}{(\delta+\rho)\left[(\delta+\rho)X_{\max} + n\zeta\right]}, \tag{6.13}$$

with $k_i^* > 0 > \lambda_{ij}^*$ everywhere. The state linearity of the game also implies that $k_i^*$ and $\lambda_{ij}^*$ also solve $\dot{k}_i = 0$ and $\dot{\lambda}_{ij} = 0$, and the same holds for $\lambda_{ii}^*$ (originating from (6.9)) and (6.6).

At this point, we may impose symmetry upon all variables, thereby dropping indexes. It is evident that the optimal advertising effort is linearly increasing in $\beta$, as intuition would suggest a priori:

$$\frac{\partial k^*}{\partial \beta} = \frac{P\beta\,(n-1)}{2b\,(\delta+\rho)\,[(\delta+\rho)\,X_{\max}+n\zeta]} > 0. \tag{6.14}$$

Less obvious is the interpretation of the following partial derivatives:

$$\begin{aligned} \frac{\partial k^*}{\partial \zeta} &= -\frac{P\,[1-\beta\,(n-1)]\,X_{\max}}{2b\,[(\delta+\rho)\,X_{\max}+n\zeta]^2} < 0 \\ \frac{\partial^2 k^*}{\partial \zeta^2} &= \frac{nP\,[1-\beta\,(n-1)]\,X_{\max}}{b\,[(\delta+\rho)\,X_{\max}+n\zeta]^3} > 0, \end{aligned} \tag{6.15}$$

which have opposite signs. This tells that the equilibrium individual effort is decreasing and convex in $\zeta$ for all $\beta \in [0, 1/(n-1))$, becoming insensitive to the natural growth rate in correspondence of the upper bound of the parameter scaling the negative advertising spillover.

**Remark 6.2** The individually optimal advertising effort increases in $\beta$ while being monotonically decreasing in $\zeta$.

The above remark prompts for the analysis of the marginal rate of substitution between $\beta$ and $\zeta$, by looking at the total differential of $k^*$ in the space $(\beta, \zeta)$, whereby $k^*$ is constant provided that

$$\frac{\partial k^*}{\partial \beta} d\beta + \frac{\partial k^*}{\partial \zeta} d\zeta = 0. \tag{6.16}$$

Its solution,

$$\frac{\partial k^*/\partial \beta}{\partial k^*/\partial \zeta} = \frac{(\delta+\rho)\,[1+\beta\,(n-1)]\,X_{\max}}{(n-1)\,\zeta\,[(\delta+\rho)\,X_{\max}+n\zeta]} \tag{6.17}$$

is positive everywhere. This yields

**Corollary 6.3** *Parameters $\beta$ and $\zeta$ are complements along any isoquant along which $k^*$ is constant.*

We are also interested in evaluating the impact of firms' number on individual and aggregate advertising efforts. To evaluate this aspect, we may define $K^* = nk^*$ and look at

$$
\begin{aligned}
\frac{\partial k^*}{\partial n} &= \frac{P\zeta\left[\beta\left(\delta+\rho\right)X_{\max}+\zeta\left(1+\beta\right)\right]}{2b\left(\delta+\rho\right)\left[\left(\delta+\rho\right)X_{\max}+n\zeta\right]^2} > 0 \\
\frac{\partial^2 k^*}{\partial n^2} &= -\frac{n\zeta^2\left[\beta\left(\delta+\rho\right)X_{\max}+\zeta\left(1+\beta\right)\right]}{b\left(\delta+\rho\right)\left[\left(\delta+\rho\right)X_{\max}+n\zeta\right]^3} < 0
\end{aligned}
\tag{6.18}
$$

and

$$
\begin{aligned}
\frac{\partial K^*}{\partial n} &= k^* + n \cdot \frac{\partial k^*}{\partial n} > 0 \\
\frac{\partial^2 K^*}{\partial n^2} &= \frac{P\zeta X_{\max}\left[\beta\left(\delta+\rho\right)X_{\max}+\zeta\left(1+\beta\right)\right]}{b\left[\left(\delta+\rho\right)X_{\max}+n\zeta\right]^3} > 0,
\end{aligned}
\tag{6.19}
$$

which can be summarized in

**Remark 6.4** All else equal, any increase in the number of firms induces an increase in individual and aggregate advertising efforts.

This result has a neatly Arrovian flavor (Arrow, 1962), as increasing market fragmentation increases aggregate investment. Here, we are dealing with advertising campaigns, while in the diachronic debate between Arrow 1962 and Schumpeter 1942 the subject was technological innovation, with Schumpeter claiming that the industry structure endowed with the highest innovation incentives should be pure monopoly, and Arrow advocating exactly the opposite.[4] Of course, this finding shall not be taken literally, as efforts and therefore costs shooting up to infinity are inadmissible as they would drive profits below zero well before that.

Before addressing this issue, we must characterize the steady-state solution, by inserting $k^*$ into (6.3), which under symmetry becomes

$$
\dot{x} = \zeta\left(1-\frac{nx}{X_{\max}}\right) + k^*\left[1-\beta\left(n-1\right)\right] - \delta x, \tag{6.20}
$$

and impose stationarity to obtain the following expression, measuring steady-state individual sales:

$$
x^* = \frac{\left[\zeta + k^*\left(1-\beta\left(n-1\right)\right)\right]X_{\max}}{\left(\delta+\rho\right)X_{\max}+n\zeta}, \tag{6.21}
$$

which is always positive under Assumption A. Moreover, it can be easily checked that Assumption B ensures that $nx^*/X_{\max} < 1$.

[4] For exhaustive overviews of the ensuing literature, still very lively, see Tirole (1988), Reinganum (1989), Martin (1993), Gilbert (2006), and Aghion et al. (2015), among others.

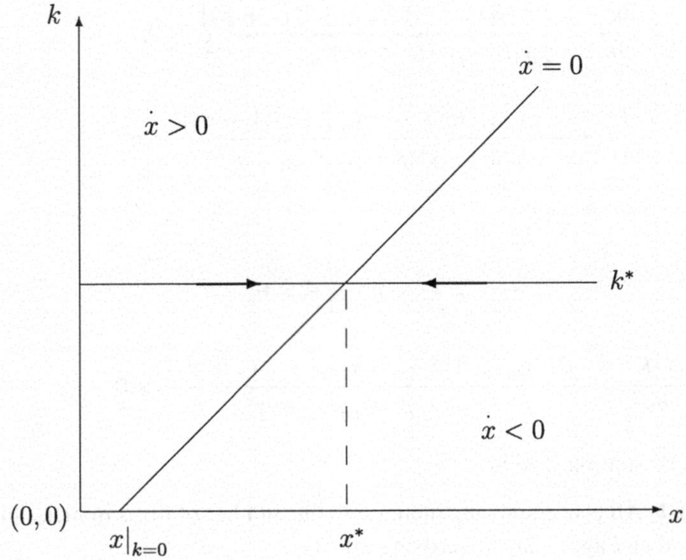


**Fig. 6.1** The phase diagram

It is now time to deal with the phase diagram and the associated stability properties in the state-control space. The equation of the steady-state locus $\dot{x} = 0$ is

$$k_{ss} = \frac{x_{ss}\,(n\zeta + \delta X_{\max}) - \zeta X_{\max}}{(1 - \beta\,(n-1))\,X_{\max}}, \tag{6.22}$$

which appears in Fig. 6.1, together with the flat line identifying the optimal advertising control $k^*$ at any point in time and for any admissible state.

The steady-state locus departs from $x|_{k=0} = \zeta X / (n\zeta + \delta X_{\max})$, i.e., the level dictated by the intrinsic properties of sales dynamics, neither boosted nor diminished by any advertising campaigns.

The sign of $\dot{x}$, which is explicitly indicated and summarized by the arrows along the flat optimal control, reveals that the steady state is stable. Indeed, the inspection of the Jacobian matrix and its determinant reveals that the open-loop equilibrium is a saddle point. The Jacobian matrix is

$$J = \begin{bmatrix} \dfrac{\partial \dot{x}}{\partial x} = -\dfrac{n\zeta + \delta X_{\max}}{\zeta X_{\max}} & \dfrac{\partial \dot{x}}{\partial k} = 1 - \beta\,(n-1) \\[2ex] \dfrac{\partial \dot{k}}{\partial x} = 0 & \dfrac{\partial \dot{k}}{\partial k} = \delta + \rho + \dfrac{[1 - \beta\,(n-1)]\,\xi}{X_{\max}} \end{bmatrix}. \tag{6.23}$$

Since the optimal control is independent of the state at all times, then obviously $\partial \dot{k} / \partial x = 0$, which in turn implies that the determinant of the Jacobian matrix collapses to the product along the main diagonal:

$$\Delta (J) = \frac{\partial \dot{x}}{\partial x} \cdot \frac{\partial \dot{k}}{\partial k} = -\frac{n\zeta + \delta X_{\max}}{\zeta X_{\max}} \left[ \delta + \rho + \frac{(1 - \beta (n-1)) \xi}{X_{\max}} \right]. \tag{6.24}$$

The expression on the r.h.s. of (6.24) is negative in view of Assumption A. This suffices to claim

**Proposition 6.5** *The steady-state point* $(x^*, k^*)$ *is a saddle point equilibrium.*

## 6.4 Concluding Remarks

As already stressed in the foregoing discussion, there have been frequent and relevant intersections between the search for strongly time consistent open-loop equilibria and the formulation of the multiple strands of the literature discussing differential games of advertising.

The frequent presence of an explicit non-separability between controls and states in dynamic games of advertising has implied, more often than not, the impossibility of characterizing feedback equilibria, confining attention to open-loop or closed-loop memoryless ones. Therefore, any such games paving the way to a (possibly degenerate) feedback solution under open-loop information have been intensively sought after.

With this in mind, we have proposed a plausible reformulation of the sales expansion model *à la* Vidale and Wolfe (1957) appearing in Deal (1979), transforming the original multiplicative interplay between controls and states into an additive one, so as to make the resulting optimal control delivered by the open-loop solution strongly time consistent without overturning the essential economic features of the model. In addition to preserving the saddle point stability of the resulting equilibrium, this approach has also allowed us to identify a well-defined Arrovian nature of the optimal industry investment in advertising campaigns under an admittedly simple and yet robust feedback rule.

## References

Aghion, P., Akcigit, U., & Howitt, P. (2015). The schumpeterian growth paradigm. *Annual Review of Economics, 7*, 557–575.

Arrow, K. (1962). Economic welfare and the allocation of resources for invention. In R. Nelson (Ed.), *The rate and direction of inventive activity*. Princeton University Press.

Bass, F. M. (1969). New product growth model for consumer durables. *Management Science, 15*, 215–227.
Becker, G., & Murphy, K. (1993). Simple theory of advertising as a good or bad. *Quarterly Journal of Economics, 108*, 941–964.
Case, J. (1979). *Economics and the competitive process*. New York University Press.
Cellini, R., & Lambertini, L. (2003a). Advertising in a differential oligopoly game. *Journal of Optimization Theory and Applications, 116*, 61–81.
Cellini, R., & Lambertini, L. (2003b). Advertising with spillover effects in a differential oligopoly game with differentiated goods. *Central European Journal of Operations Research, 11*, 409–423.
Cellini, R., Lambertini, L., & Mantovani, A. (2008). Persuasive advertising under bertrand competition: A differential game. *Operations Research Letters, 36*, 381–314.
Clark, C. W. (1990). *Mathematical bioeconomics: The optimal management of renewable resources*. Wiley.
Clemhout, S., Leitmann, G., & Wan, Jr. H. Y. (1971). A differential game model of duopoly. *Econometrica, 39*, 911–938.
d'Aspremont, C., & Jacquemin, A. (1988). Cooperative and noncooperative R&D in duopoly with spillovers. *American Economic Review, 78*, 1133–1137.
Deal, K. (1979). Optimizing advertising expenditure in a dynamic duopoly. *Operations Research, 27*, 682–692.
Dockner, E.J, Jørgensen, S., Long, N. V., & Sorger, G. (2000). *Differential games in economics and management science*. Cambridge University Press.
Dragone, D., Lambertini, L., & Palestini, A. (2010). The Leitmann-Schmitendorf advertising game with n players and time discounting. *Applied Mathematics and Computation, 217*, 1010–1016.
El Ouardighi, F., Jørgensen, S., & Pasin, F. (2013). A dynamic game with monopolist manufacturer and price-competing duopolist retailers. *OR Spectrum, 35*, 1059–1084.
Erickson, G. (2003). *Dynamic models of advertising competition* (2nd ed.). Kluwer.
Feichtinger, G. (1983). The nash solution of an advertising differential game: Generalization of a model by Leitmann and Schmitendorf. *IEEE Transactions on Automatic Control, 28*, 1044–1048.
Fershtman, C. (1984). Goodwill and market shares in oligopoly. *Economica, 51*, 271–281.
Friedman, L. (1958). Game-theory models in the allocation of advertising expenditures. *Operations Research, 6*, 699–709.
Garvin, D. (1988). *Managing quality*. Free Press.
Gilbert, R. (2006). Looking for Mr Schumpeter: Where are we in the competition-innovation debate? In J. Lerner & S. Stern (Eds.). *Innovation policy and economy*. NBER, MIT Press.
Gould, J. P. (1970). Diffusion processes and optimal advertising policy. In E. S. Phelps (Ed.), *Microeconomic foundations of employment and inflation theory* (pp. 338–368). Norton.
Jørgensen, S., Martín-Herrán, G., & Zaccour, G. (2010). The Leitmann-Schmitendorf advertising differential game. *Applied Mathematics and Computation, 217*, 1110–1116.
Jørgensen, S., & Zaccour, G. (2004). *Differential games in marketing*. Kluwer.
Lambertini, L. (2018). *Differential games in industrial economics*. Cambridge University Press.
Lambertini, L., & Leitmann, G. (2019). On the attainment of the maximum sustainable yield in the Verhulst-Lotka-Volterra model. *Automatica, 110*, article 108555, 1–5.
Lambertini, L., & Zaccour, G. (2015). Inverted-U aggregate investment curves in a dynamic game of advertising. *Economics Letters, 132*, 34–38.
Lanchester, F. W. (1956). Mathematics in warfare. In J. R. Newman (Ed.), *The world of mathematics*. Simon and Schuster.
Leitmann, G., & Schmitendorf, W. E. (1978). Profit maximization through advertising: A nonzero sum differential game approach. *IEEE Transactions on Automatic Control, 23*, 646–650.
Martin, S. (1993). *Advanced industrial economics*. Blackwell.
Nerlove, M., & Arrow, K. (1962). Optimal advertising policy under dynamic conditions. *Economica, 29*, 129–142.

Reinganum, J. (1989). The timing of innovation: Research, development and diffusion. In R. Schmalensee & R. Willig (Eds.), *Handbook of industrial organization*. North-Holland.

Schumpeter, J. (1942). *Capitalism, socialism and democracy*. Allen & Unwin.

Sethi, S. (1983). Deterministic and stochastic optimization of a dynamic advertising model. *Optimal Control Applications and Methods, 4*, 179–184.

Sethi, S., & Thompson, G. (1981). *Optimal control theory: Applications to management science*. Nijhoff.

Sorger, G. (1989). Competitive dynamic advertising: A modification of the case game. *Journal of Economic Dynamics and Control, 13*, 55–80.

Stigler, G., & Becker, G. (1977). De Gustibus non Est Disputandum. *American Economic Review, 67*, 7–90.

Tirole, J. (1988). *The theory of industrial organization*. MIT Press.

Vidale, M., & Wolfe, H. (1957). An operations research study of sales response to advertising. *Operations Research, 5*, 370–381.

# Chapter 7
# The Cost of Myopia with Respect to a Switching Time in an Advertising Model

**Alessandra Buratto, Luca Grosset, Maddalena Muttoni, and Bruno Viscolani**

**Abstract** The ability to react to abrupt changes constitutes a fundamental skill for decision makers, especially in dynamic contexts where problem structures can change over time. However, there are situations in which planners are myopic, i.e., unaware of the impending changeover. The latter context inevitably results in a loss of profit.

This paper aims to assess the cost of adopting a myopic approach toward system changes. We consider a marketing problem modeled à la Nerlove and Arrow, where the demand for a product is influenced by the goodwill of the firm that produces, advertises, and sells it. Moreover, we assume that production costs may change abruptly with a hazard rate that depends on the demand for the product.

To address this situation, we formulate and solve an optimal control problem with stochastic switching time. We compare the optimal profit of a planner who is aware of the possibility of a switch to the one of a planner who is myopic with respect to such an event.

**Keywords** Optimal control · Stochastic switching · Myopic decision-making · Dynamic advertising model · Abrupt system changes

## 7.1 Introduction

Advertising is one of the most effective marketing tools that can influence consumer behavior, leading to changes in demand for a product or a service. There are two primary models in the literature on optimal control applications to advertising, both proposed around the same time. The first model, by Vidale and Wolfe (1957), describes the response of sales to advertising and aims to represent typical behaviors observed in real data. The second model, by Nerlove and Arrow (1962),

A. Buratto (✉) · L. Grosset · M. Muttoni · B. Viscolani
Department of Mathematics "Tullio Levi-Civita", University of Padova, Padova, Italy
e-mail: buratto@math.unipd.it



F. El Ouardighi, G. Feichtinger (eds.), *The Unaffordable Price of Static Decision-making Models*, International Series in Operations Research & Management Science 365, https://doi.org/10.1007/978-3-031-88638-6_7

assumes that the demand and sale intensity of a product depend on a state variable called goodwill, which represents the effects of a firm's investment in advertising. The Nerlove-Arrow model has become an essential reference for advertising and marketing research, as seen in the review articles Sethi (1977), Feichtinger et al. (1994), and Huang et al. (2012). From this seminal model, we analyze how the optimal advertising campaign adapts in anticipation of a regime shift. In more detail, we consider an advertising problem for a firm assuming that its production costs can disruptively change during the programming interval and affect its (marginal) profits. We assume that the time at which the switch occurs is affected by the demand due to the concept of economies or diseconomies of scale. While it's commonly expected that production costs will decrease as production volume increases (economies of scale), there are situations where the opposite occurs due to various challenges that arise with growth and increased demand. For example, increased demand can lead to raw material shortages, and therefore suppliers might struggle to meet the higher demand. Additionally, a growing demand could lead to labor shortages, particularly if specialized skills are required. This can result in companies needing to offer higher wages or overtime pay to attract workers. In both cases, as a result, production costs can increase abruptly.

The stochastic time corresponding to an increase in production costs can be modeled as a random variable, named the switching time, whose distribution is influenced by the state variable of the system.

Reacting to sudden changes is an important skill for decision makers. Strategic planning and a farsighted perspective are crucial to managing the potential risks associated with irreversible changes in production costs.

In this paper, we compare two different types of behavior, assuming that in any case, the entire advertising campaign must be planned at the beginning of the programming interval. In the first case, we assume that the firm has complete information about the switching time and can plan how to adjust its advertising campaign for any occurrence of the switching time if such a switch occurs during the programming interval.

In the second case, we analyze a firm that has no information on the time of the switch and plans its advertising strategy as if nothing would change at all.

We denote the latter firm as *myopic with respect to the switching time* (in short, *myopic*). In the literature, the former type of behavior is called *anticipative* (Buratto et al., 2022); nevertheless, for a more immediate distinction between the two types of planners, in this paper we shall refer to it as *non-myopic*.

Within the attitude of the myopic firm's with respect to the switching time, we further distinguish two scenarios. First, the decision maker is unable to update its control if the switching time is realized. Due to initial agreements, its advertising campaign will remain as fixed at the initial time. Alternatively, in a second scenario, the decision maker, although myopic, can adapt their strategy to the situation that arises after the switching time.

With our model, we want to analyze the following three research questions:

- How do optimal advertising policies and expected profits vary for the two types of decision maker?
- What is the cost of myopia? It can be quantified by examining the decrease in expected profit, which is directly related to the decrease in the level of knowledge of the decision maker about the time of the switch.
- While the expected profit of the myopic planner is lower than that of the non-myopic planner, there are certain instances where, due to the specific realization of the switching time (e.g., in our model, if the change occurs later on), the actual profit of the myopic planner may be higher than that of the farsighted planner. Therefore, our third research question is: With what probability does the myopic decision maker achieve a higher profit?

This work is organized as follows. In Sect. 7.2, we present a marketing scenario based on the Nerlove-Arrow framework and model its fundamental features, particularly focusing on switching time, to describe a disruptive change in the firm's production costs. In Sect. 7.3, using the necessary conditions for optimal control of heterogeneous systems (Veliov, 2008), we present the necessary conditions for a non-myopic decision maker and find the optimal advertising campaign up to integration of the state-adjoint system of ODEs that is fully nonlinear. In Sect. 7.4, we present the necessary conditions for a myopic decision maker and we find the optimal advertising campaign. In Sect. 7.5, we numerically compare the optimal advertising campaign and the optimal expected profits. In the Conclusions, we describe some open questions connected with this advertising model.

## 7.2 Model

We consider a finite-time marketing problem in which a company invests in advertising at a rate $a(t) \geq 0$ to increase the demand for its product. The time horizon $[0, T]$ is finite (with $T > 0$), allowing us to set a constant selling price for the product. In fact, we are assuming that during the programming interval, the price remains constant and cannot be modified by the firm. Following the Nerlove and Arrow model, we assume that the *goodwill* summarizes the effect of advertising investment; hence, $G(t)$ is a state variable which increases with the firm's advertising intensity $a(t)$, and it decays exponentially at a constant rate $\delta > 0$ if not sustained by the advertising:

$$\begin{cases} \dot{G}(t) = a(t) - \delta G(t) & \text{for } t \in [0, T] \\ G(0) = G_0 > 0 \end{cases} \tag{7.1}$$

The firm's objective is to maximize its payoff, which is composed of an intertemporal term and a salvage value. The intertemporal term captures the trade-off between the profit from selling the product and the cost of promoting it, whereas the salvage value captures the interest of the firm in sustaining the brand value. The unit

**Table 7.1** Variables and parameters

| | |
|---|---|
| $a(t) \geq 0$ | Advertising investment at time $t$ (control function) |
| $G(t)$ | Goodwill level at time $t$ (state function) |
| $\alpha, \beta$ | Demand parameters, $\alpha, \beta > 0$ |
| $c > 0$ | Unit production cost |
| $p > c$ | Unit selling price |
| $\kappa > 0$ | Parameter for quadratic advertising cost |
| $\sigma > 0$ | Marginal weight of the final goodwill |
| $\delta > 0$ | Goodwill's depreciation rate |

production cost $c > 0$ is constant. As mentioned above, we also assume that the unit selling price $p > c$ is constant throughout the programming interval. Let us assume that instantaneous demand depends linearly on the goodwill's value, according to the following formula ($\alpha, \beta > 0$):

$$D(G) = \alpha + \beta G \tag{7.2}$$

Hence, the instantaneous firm's profit from selling the product is $(p-c)D(G(t)) = (p-c)(\alpha + \beta G(t))$. The advertising cost is assumed to be quadratic, which is a standard hypothesis in the related literature, and we denote by $\kappa > 0$ the advertising cost parameter. The salvage value is assumed to be proportional to the final goodwill $G(T)$, with weight $\sigma > 0$. This parameter allows the firm to maximize its brand value even at the end of the programming interval. Summarizing, the firm wants to solve the following optimal control problem:

$$\underset{a(t)\geq 0}{\text{maximize}} \int_0^T \left[(p-c)(\alpha + \beta G(t)) - \frac{\kappa}{2}a^2(t)\right] dt + \sigma G(T) \tag{7.3}$$

subject to (7.1):

$$\begin{cases} \dot{G}(t) = a(t) - \delta G(t) \\ G(0) = G_0 \end{cases}$$

A summary of all notations used in the chapter is included in Table 7.1

### 7.2.1 Stochastic Switching Time: Rise in Production Costs

A stochastic switching time in a dynamic system is an unpredictable event that occurs at a stochastic time $\tau$ that abruptly changes the nature of the problem. The instant $\tau$ can be modeled as an absolutely continuous random variable with support $[0, +\infty)$. Denoting by Stage 1 the period before the switch (i.e., all $t \leq \tau$) and Stage 2 the period after the switch (i.e., all $t > \tau$), we have $c_1 < p$ as a unit production cost in Stage 1, while $c_2 < p$ in Stage 2, with the assumption that $c_1 < c_2$.

In our model, we assume such a change to be also irreversible and the instant $\tau$ represents a sudden rise in the production cost, which in turn can be formalized as follows:

$$c = \begin{cases} c_1 & \text{in Stage 1} \\ c_2 & \text{in Stage 2} \end{cases} \tag{7.4}$$

The literature on problems related to switching time is wide and many articles tackle the problem of reacting to abrupt changes. A well-established and widely used method to solve this kind of optimal control problems is given by the *backward approach*, a particular case of the more general theory of piecewise deterministic optimal control problems (see Dockner et al., 2000, Ch. 8.1), where the system switches between "modes" at stochastic times, but the dynamics and running payoff in each mode are deterministic. In the backward approach, see, e.g., Boukas et al. (1990), the Stage 2 value function acts as a salvage value for the Stage 1 problem, and the random switching time constitutes a random endpoint for Stage 1.

A different example on an infinite horizon, in Tsur and Zemel (2017), assumes the existence of multiple (although with identical hazard rates and effects) catastrophic threats with state-dependent hazards. The authors, Tsur and Zemel, are very prolific in the stream of literature on two-stage optimal control with stochastic switching time, mostly with an infinite horizon, with a state-dependent hazard rate, and with a backward approach.

A parallel literature substream, shared by this paper, concerns the same type of problem, tackled with a new solution method, here called *heterogeneous approach*. This method entails formulating an equivalent deterministic optimal control problem, distributed along an additional variable that represents the occurrence of the switching time. Wang (1964) and Brogan (1968) employ a dynamic programming approach to derive the MP, enabling the interpretation of adjoint variables as shadow prices. Feichtinger et al. in Feichtinger et al. (2003) obtain a global maximum principle to tackle this kind of problem, while Veliov in Veliov (2008) provides the necessary conditions for the solution of a more general heterogeneous optimal control problem, which can be applied to this particular case.

More recently, in Wrzaczek et al. (2020), Wrzaczek et al. describe the transformation of the two-stage problem into a heterogeneous one, discuss the advantages of this approach compared to the standard backward approach, and provide a simple example on a macroeconomic shock with state-dependent hazard rate in infinite horizon. The same approach is used by Kuhn and Wrzaczek in Kuhn and Wrzaczek (2021), for an infinite-time, two-stage rational addiction model that explicitly incorporates a pre-addiction phase and a stochastic transition into addiction with a state-dependent hazard. In the same substream, Buratto et al. (2022) features an infinite-time, two-stage SIR model with lockdown measures and R&D, where the stochastic switch represents the discovery of an effective vaccine. To this day, this is the only published work in which the hazard rate is directly controllable (depending on time and R&D effort). A recent comprehensive overview of dynamic economic problems with regime switches can be found in Haunschmied et al. (2021). Finally, in Freiberger (2023), the author provides a Julia package for solving two-stage optimal control problems with random switching times. This work applies necessary optimality conditions for age-structured optimal control models to a case study on the health impacts of air pollution.

Our work fits into the latter heterogeneous approach substream; however, differently from all the works discussed above, we consider a model in a finite time horizon.

A classical way to tackle optimal control problems with stochastic switching time is to introduce a function called *hazard rate,* or *switching rate*, which describes the probability of the occurrence of such a switch. The hazard rate may be exogenous or endogenous and in the latter case it may depend both on the state variable and the control variable (as, e.g., in Sorger, 1991 and in Dawid et al., 2015). In this model, we assume the hazard rate to be endogenous and dependent on the demand of the product $D(G)$, in (7.2); therefore, it only depends on the state $G(\cdot)$ of the system.

We assume that the absolutely continuous random variable $\tau$ is defined by the following equation:

$$\lim_{h \to 0^+} \frac{\mathbb{P}(\tau \leq t + h \,|\, \tau > t)}{h} = \eta\big(D(G(t))\big) \tag{7.5}$$

where $\eta : (0, +\infty) \to (0, +\infty)$ is the *hazard rate* function. The distribution of $\tau$ can be derived from the definition of $\eta$; however, since the goodwill function $G(t)$ is not defined after $T$, we can determine the distribution of $\tau$ only within the programming interval. Nevertheless, we are not interested in the distribution of $\tau$ after $T$: It suffices to know that it can be extended in any way so that the total integral of $\tau$'s probability density equals 1 in $[0, +\infty)$.

In the following, we assume that the hazard rate is a linear and increasing function of the demand

$$\eta(D) = \varepsilon D = \varepsilon(\alpha + \beta G), \qquad \varepsilon > 0, \tag{7.6}$$

where $\varepsilon$ represents the marginal hazard with respect to the demand.

Since we consider a finite time horizon, $\tau$ could occur during the programming interval or later. If it occurs before the final time $T$, it splits the planning horizon into a Stage 1 and a Stage 2, respectively, before and after $\tau$; if it occurs after $T$, the entire planning horizon is covered by Stage 1. The latter case implies that, with positive probability, the unit production price can be equal to $c_1$ throughout the whole programming interval.

### *7.2.2 Switching Time: Information and Adaptability*

In this work, we study the key role that information plays in planning the optimal strategy under the uncertainty of a pending switching time. Our aim is to analyze how the ability to adapt strategies based on the available level of information allows to increase profits, emphasizing the importance of dynamic methods for accurate managerial prescriptions. We do so by considering two types of decision makers.

The first type of decision maker is familiar with both the goodwill's dynamics and the influence of their control on the evolution of the probability distribution of the switching time. In addition, they anticipate the effects that the switch will have on the system. This leads them to define, for Stage 1, a control that covers the entire programming interval, because they do not know, a priori, when the switch will occur. This control will be truncated as soon as the switch occurs and Stage 2 starts. Hereafter, we will denote by *non-myopic* a decision maker belonging to this first type.

Concerning Stage 2, non-myopic decision makers plan, at the beginning of the programming interval, a strategy that adapts to the realization of the random variable, so that they determine a family of controls, parameterized by the realization of the switching time, each of them defined in the time interval after the switch. In other words, the Stage 2 control for $\tau = s$ is defined in the Stage 2 interval $[s, T]$.

Let us delve into the details, trying to establish the mathematical framework that allows us to describe this problem.

- Stage 1: the planner is expecting $\tau$ to occur at any time and knows its hazard rate function $\eta(D(G(t)))$ and the effect it will have on the system, that is, the future production cost $c_2$. Therefore, they will have to balance the increase in goodwill and consequently in demand with the probability of sudden increases in production costs. Since the planner does not know exactly when $\tau$ will occur, their Stage 1 advertising strategy needs to cover the whole programming interval $[0, T]$; the Stage 1 process will be described by the following couple of functions:

$$(a_1(t),\ G_1(t)) \quad \text{for } t \in [0, T]$$

- Stage 2: the planner notices when $\tau$ occurs and can update their strategy according to the new regime in the interval $[\tau, T]$. We assume that the decision maker establishes a parametric control function at the beginning of the process. Since different realizations of $\tau$ may lead to different optimal Stage 2 strategies, the Stage 2 process will actually be parametrized by the realization $s$ of the switching time $\tau$ during the planning horizon:

$$(a_2(s,t),\ G_2(s,t)), \quad \text{for } (s,t) \in \Delta := \{(s,t) \mid s \in [0, T],\ t \in [s, T]\}$$

The firm plans their strategy ahead for both stages and for every possible occurrence of $\tau$: before the programming interval starts, they will have decided both $a_1(t)$ and $a_2(s, t)$. If $\tau$ occurs at time $s$, the firm will implement the strategy $a_1(t)$ for $t \in [0, s]$, and then the strategy $a_2(s, \cdot) : [s, T] \to [0, +\infty)$.

**Remark 7.1** We emphasize that throughout the remainder of this paper we will use this notation for the state and control functions of Stage 2: the first variable $s$

represents the realization of the switching time, while the second variable $t \in [s, T]$ represents the time.

A second type of decision maker is familiar with the goodwill's dynamics but ignores the possibility of a switching time. They choose an advertising strategy that covers the entire programming interval $[0, T]$. This strategy is fixed at the beginning of the programming interval. The control and state functions are then represented by

$$(a(t), G(t)), \quad \text{for } t \in [0, T].$$

In what follows, we will denote *myopic* this second type of decision maker.

It is worth observing that even though myopic decision makers do not consider the switching time, such an event can indeed occur. Therefore, recalling that our model is based on the assumption that the hazard rate depends on the demand for the product, myopic decision makers determine their strategies without knowing that their controls influence the distribution of the random variable $\tau$. To properly compare the two types of decision makers, we will calculate the expected profit both for non-myopic and myopic decision makers, although taking into account that the latter are unaware of the randomness of the system. Indeed, the non-myopic planners have all the information about the system's stochasticity, so they are able to compute the optimal strategy to maximize the expected payoff over all possible realizations of the switching time. On the other hand, myopic planners maximize their profit as if no switch were to occur; however, even if they do not consider it, their actions do influence the switch's hazard rate, and their production cost will indeed increase at some point. Knowing this, we can evaluate the *actual* expected profit of myopic planners, which will necessarily differ from their objective value.

For simplicity, it is convenient to perform the computation of the profit first for the non-myopic decision maker. We will see that it is possible to treat the myopic case as a formal instance of the non-myopic case.

## 7.3 Non-myopic Decision Maker

Starting with the benchmark model (7.3), in order to formalize the switching time optimal control problem for the non-myopic decision maker, we need to introduce some notation. Planners who are aware that a change in marginal costs can occur aim to maximize their expected profit, in the set of feasible control paths $a(\cdot)$, so that the density probability $\mathbb{E}_{a(\cdot)}$ is needed. More precisely, since it is the control used in Stage 1 that modifies the state variable and, in turn, the distribution function of the random variable $\tau$, this dependence must be indicated on the expected value operator. The formulation of the problem has to take into account both the possibility that the switch occurs before the final time $T$ of the programming period or after it. With such an attempt, let us introduce the indicator functions $\mathbb{1}_{\{\tau<T\}}$ and $\mathbb{1}_{\{\tau\geq T\}}$,

respectively. In the first case, there will be a Stage 1, with a payoff equal to (7.3) and a Stage 2 with a payoff with a new marginal cost $c_2$. On the other hand, in the latter case, i.e., if the switch occurs after the end of the programming period, nothing will change, and the problem essentially remains equal to the one in (7.3) with the original marginal cost $c_1$. Finally, the problem of the non-myopic decision maker can be stated as follows:

$$\begin{aligned}\underset{a_1(t),a_2(s,t)\geq 0}{\text{maximize}}\ \mathbb{E}_{a_1(t)}\Bigg[\mathbb{1}_{\{\tau<T\}}\Bigg\{\int_0^{\tau}\Big[(p-c_1)D(G_1(t))-\frac{\kappa}{2}a_1(t)^2\Big]dt\\+\int_{\tau}^{T}\Big[(p-c_2)D(G_2(\tau,t))-\frac{\kappa}{2}a_2(\tau,t)^2\Big]dt+\sigma G_2(\tau,T)\Bigg\}\\+\mathbb{1}_{\{\tau\geq T\}}\Bigg\{\int_0^{T}\Big[(p-c_1)D(G_1(t))-\frac{\kappa}{2}a_1(t)^2\Big]dt+\sigma G_1(T)\Bigg\}\Bigg]\end{aligned}\tag{7.7}$$

subject to:

$$\begin{cases}\dot{G}_1(t)=a_1(t)-\delta G_1(t), & \text{for } t\in[0,\tau)\\ G_1(0)=G_0 & \\ \dot{G}_2(\tau,t)=a_2(\tau,t)-\delta G_2(\tau,t), & \text{for } t\in[\tau,T]\\ G_2(\tau,\tau)=G_1(\tau) & \end{cases}\tag{7.8}$$

where the hazard rate of $\tau$ is defined in (7.5) and, with a common abuse of notation, see, e.g., Wrzaczek et al. (2020), we have written $\dot{G}_2(s,t)=\partial_t G_2(s,t)$.

We emphasize that the probability law that governs the process also depends on the chosen control $a(t)$, and henceforth, in accordance with Sorger (1991) and (Dockner et al., 2000, p.204), we write $\mathbb{E}_{a(t)}$ to denote the expectations computed with respect to that law.

To study this problem, we need to compute the expectation through the probability density of $\tau$. For this purpose, it is convenient to introduce an auxiliary state variable $z_1(t)=\mathbb{P}(\tau>t)$ for Stage 1. It represents the probability of still being in Stage 1 at time $t$. This definition allows us to write the probability density of $\tau$, which is the derivative of $\mathbb{P}(\tau\leq t)=1-\mathbb{P}(\tau>t)$, as:

$$f_\tau(t)=-\dot{z}_1(t).\tag{7.9}$$

Following the same computation performed in Wrzaczek et al. (2020), we can prove that $z_1(t)$ is the solution of the following Cauchy problem:

$$\begin{cases}\dot{z}_1(t)=-\eta(D(G_1(t))z_1(t), & \text{for } t\in[0,T],\\ z_1(0)=1 & \end{cases}\tag{7.10}$$

These two results allow us to write explicitly the expected value introduced in the objective functional. After basic integral manipulation, the firm's objective functional becomes:

$$\int_0^T z_1(t)\big[(p-c_1)D(G_1(t)) - \frac{\kappa}{2}a_1(t)^2\big]\,dt + z_1(T)\sigma G_1(T)$$
$$+ \int_0^T \eta\big(G_1(s)\big)z_1(s)\Big\{\int_s^T \big[(p-c_2)D(G_2(s,t))$$
$$- \frac{\kappa}{2}a_2(s,t)^2\big]\,dt + \sigma G_2(s,T)\Big\}\,ds$$

It is worth observing how the auxiliary state variable $z_1(t)$ acts as a discount factor for the Stage 1 payoff. In order to be able to treat this maximization problem with the theory provided in Veliov (2008), we first need to separate the payoff into two additive terms containing Stage 1 and Stage 2 variables. The problem with the above formulation is that the Stage 2 payoff (starting from $s$) is multiplied by $\eta\big(G_1(s)\big)z_1(s)$, which depends on the Stage 1 variables $G_1$ and $z_1$. We work around this as in Wrzaczek et al. (2020), by introducing the auxiliary Stage 2 variable $z_2(s,t) = \eta\big(G_1(s)\big)z_1(s)$, i.e.,

$$\begin{cases} \dot{z}_2(s,t) = 0 \\ z_2(s,s) = \eta\big(D(G_1(s))\big)z_1(s) \end{cases} \tag{7.11}$$

The variable $z_2(s,t)$ depends on the switching time $s$ and it is constant in time $t$. It represents the probability density of $\tau$ at time $s$. After substituting $z_2$ in the objective functional, we obtain

$$\int_0^T z_1(t)\big[(p-c_1)D(G_1(t)) - \frac{\kappa}{2}a_1(t)^2\big]\,dt + z_1(T)\sigma G_1(T)$$
$$+ \int_0^T \Big\{\int_s^T z_2(s,t)\big[(p-c_2)D(G_2(s,t)) - \frac{\kappa}{2}a_2(s,t)^2\big]\,dt$$
$$+ z_2(s,T)\sigma G_2(s,T)\Big\}\,ds$$

Problem (7.7) can be reformulated as a deterministic, heterogeneous one:

$$\underset{a_1(t),a_2(s,t)\geq 0}{\text{maximize}} \Big[\int_0^T z_1(t)\big[(p-c_1)D(G_1(t)) - \frac{\kappa}{2}a_1(t)^2\big]\,dt + z_1(T)\sigma G_1(T)$$
$$+ \int_0^T \Big\{\int_s^T z_2(s,t)\big[(p-c_2)D(G_2(s,t)) - \frac{\kappa}{2}a_2(s,t)^2\big]\,dt$$
$$+ z_2(s,T)\sigma G_2(s,T)\Big\}\,ds\Big] \tag{7.12}$$

subject to:

$$\begin{cases} \dot{G}_1(t) = a_1(t) - \delta G_1(t), & G_1(0) = G_0, \\ \dot{z}_1(t) = -\eta\big(D(G_1(t))\big)z_1(t), & z_1(0) = 1, \\ \dot{G}_2(s,t) = a_2(s,t) - \delta G_2(s,t), & G_2(s,s) = G_1(s), \\ \dot{z}_2(s,t) = 0, & z_2(s,s) = \eta\big(D(G_1(s))\big)z_1(s) \end{cases} \tag{7.13}$$

We have transformed the optimal control problem with stochastic switching time described at the beginning of this section into a heterogeneous deterministic optimal control problem. The idea now is to characterize its optimal solutions with necessary conditions.

**Theorem 7.1** *Let $\big(a_1^*(t), G_1^*(t), z_1^*(t), a_2^*(s,t), G_2^*(s,t), z_2^*(s,t)\big)$ be the optimal solution of the heterogeneous problem (7.12) and (7.13) for the non-myopic decision maker, then the optimal advertising efforts $a_1^*(t), a_2^*(s,t)$ in Stage 1 and Stage 2 respectively are:*

$$a_1^*(t) = \big[\lambda_G(t)/\kappa\big]^+, \qquad a_2^*(s,t) = \xi_G(s,t)/\kappa, \tag{7.14}$$

*where*

$$\xi_G(s,t) = \frac{(p-c_2)\beta}{\delta}(1 - e^{-\delta(T-t)}) + \sigma e^{-\delta(T-t)}, \quad t \geq s \tag{7.15}$$

*and the following co-state system holds:*

$$\begin{cases} \dot{\lambda}_G(t) = -(p-c_1)\beta + \delta\lambda_G(t) - \varepsilon(\alpha + \beta G_1^*(t))\big[\xi_G(t,t) - \lambda_G(t)\big] - \\ \qquad\qquad -\varepsilon\beta\big[\xi_z(t,t) - \lambda_z(t)\big] \\ \lambda_G(T) = \sigma \\ \dot{\lambda}_z(t) = -\big((p-c_1)D(G_1^*(t)) - \frac{\kappa}{2}a_1^*(t)^2\big) - \varepsilon(\alpha + \beta G_1^*(t))\big[\xi_z(t,t) - \lambda_z(t)\big] \\ \lambda_z(T) = \sigma G_1^*(T) \\ \dot{\xi}_z(s,t) = -\big((p-c_2)D(G_2^*(s,t)) - \frac{\kappa}{2}a_2^*(s,t)^2\big) \\ \xi_z(s,T) = \sigma G_2^*(s,T) \end{cases} \tag{7.16}$$

***Proof*** Let us denote by $\lambda_G$ and $\lambda_z$ the Stage 1 co-state variables, and by $\xi_G$ and $\xi_z$ the Stage 2 ones, as in Buratto et al. (2023).[1] By Veliov (2008), the maximality condition for the Stage 2 control is:

[1] For simplicity, we omit the superscript "$c$" for the current-value co-state functions that correspond to the state variable $G$.

$$a_2^*(s,t) \in \arg\max_{a \geq 0} \left\{(p - c_2)D\big(G_2^*(s,t)\big) - \frac{\kappa}{2}a^2 + \xi_G(s,t)\big[a - \delta G_2^*(s,t)\big]\right\},$$

yielding

$$a_2^*(s,t) = \big[\xi_G(s,t)/\kappa\big]^+$$

Concerning Stage 1, since the initial condition of the Stage 2 state variables and the hazard rate function do not depend on the control, the necessary condition for the control is a maximality condition of the following form (see also Buratto et al. (2023)):

$$a_1^*(t) \in \arg\max_{a \geq 0} \left\{(p - c_1)D\big(G_1^*(t)\big) - \frac{\kappa}{2}a^2 + \lambda_G(t)\big[a - \delta G_1^*(t)\big]\right\},$$

yielding

$$a_1^*(t) = \big[\lambda_G(t)/\kappa\big]^+$$

We obtain the co-state system from the more general formulation in Buratto et al. (2023), with $\phi(t, G, a) = G$, due to the continuity of the goodwill upon the switch, and hence $\partial_G \phi(t, G, a) = 1$. The co-state functions, recalling that the motion equation is the same in the two stages, satisfy the following system:

$$\begin{cases} \dot{\lambda}_G(t) = -(p - c_1)D'(G_1^*(t)) + \delta\lambda_G(t) - \eta\big(D(G_1^*(t))\big)\big[\xi_G(t,t) - \lambda_G(t)\big] - \\ \qquad\qquad - \frac{d}{dG}\big[\eta(D(G_1^*(t)))\big]\big[\xi_z(t,t) - \lambda_z(t)\big] \\ \lambda_G(T) = \sigma \\ \dot{\lambda}_z(t) = -\big((p - c_1)D(G_1^*(t)) - \frac{\kappa}{2}a_1^*(t)^2\big) - \eta\big(D(G_1^*(t))\big)\big[\xi_z(t,t) - \lambda_z(t)\big] \\ \lambda_z(T) = \sigma G_1^*(T) \\ \dot{\xi}_G(s,t) = -(p - c_2)D'(G_2^*(s,t)) + \delta\xi_G(s,t) \\ \xi_G(s,T) = \sigma \\ \dot{\xi}_z(s,t) = -\big((p - c_2)D(G_2^*(s,t)) - \frac{\kappa}{2}a_2^*(s,t)^2\big) \\ \xi_z(s,T) = \sigma G_2^*(s,T) \end{cases}$$

Observe that the Cauchy problem for $\xi_G$ can be solved independently from the other ones, obtaining equation (7.15). The remaining equations, recalling from (7.2) and (7.6) that $D'(G_1^*) = \beta$ and $\frac{d}{dG}\big[\eta(D(G_1^*))\big] = \varepsilon\beta$, constitute the co-state system (7.16).

It can be easily proved that $\xi_G(s,t) > 0$ being $\sigma > 0$, thus guaranteeing the positivity of the Stage 2 advertising effort $a_2^*(s,t)$ in (7.14). □

Since, in both stages, the optimal control depends solely on the co-state corresponding to the goodwill, it is of interest to analyze the evolution of such co-states $\lambda_G$ and $\xi_G$ (see system (7.3)) and the role played by the anticipation of the switch in shaping such evolution.

Regarding $\xi_G$, we observe that its adjoint equation (for any fixed $s$) is the same as it would be in the case of a single-stage problem with production cost equal to $c_2$. Indeed, after the switch has occurred, there is no uncertainty about future disruptions, and therefore the planner may equivalently be facing a new simple single-stage optimal control problem on the interval $[s, T]$.

As for $\lambda_G$, if we compare its adjoint equation with that in the case of a single stage problem with production cost equal to $c_1$, we notice the presence of two additional terms:

$$-\eta\big(D(G_1^*(t))\big)\big[\xi_G(t,t) - \lambda_G(t)\big] \qquad \text{and} \qquad -\varepsilon\beta\big[\xi_z(t,t) - \lambda_z(t)\big] \tag{7.17}$$

These terms represent the anticipating effect on the Stage 1 shadow value of the goodwill. Let us illustrate their meaning, starting from the latter term.

By integrating the equation for $\xi_z$, we obtain that $\xi_z(s,t)$ equals the optimal value of the Stage 2 problem starting from $t$ (with $t \geq s$), given that the switch occurred at time $s$:

$$\xi_z(s,t) = \int_t^T \big[(p - c_2)D(G_2^*(s,\theta)) - \frac{\kappa}{2}a_2^*(s,\theta)^2\big]\,d\theta + \sigma G_2^*(s,T) \tag{7.18}$$

By integrating the equation for $\lambda_z$, we obtain that $\lambda_z(t)$ equals the optimal expected value of the two-stage problem starting from $t$, given that the switch has not occurred yet at time $t$:

$$\begin{aligned}
\lambda_z(t) &= \frac{1}{z_1^*(t)}\Big\{\int_t^T z_1^*(\theta)\big[(p-c_1)D(G_1^*(\theta)) - \frac{\kappa}{2}a_1^*(\theta)^2 \\
&\qquad + \eta\big(D(G_1^*(\theta))\big)\xi_z(\theta,\theta)\big]\,d\theta + z_1^*(T)\sigma G_1^*(T)\Big\} \\
&= \frac{1}{z_1^*(t)}\Big\{\int_t^T z_1^*(\theta)\big[(p-c_1)D(G_1^*(\theta)) - \frac{\kappa}{2}a_1^*(\theta)^2 \\
&\qquad + \eta\big(D(G_1^*(\theta))\big)\Big(\int_\theta^T \big[(p-c_2)D(G_2^*(\theta,u)) - \frac{\kappa}{2}a_2^*(\theta,u)^2\big]\,du \\
&\qquad + \sigma G_2^*(\theta,T)\Big)\big]\,d\theta + z_1^*(T)\sigma G_1^*(T)\Big\} \\
&= \mathbb{E}\Big[\chi_{\tau<T}\Big(\int_t^\tau \big[(p-c_1)D(G_1^*(\theta)) - \frac{\kappa}{2}a_1^*(\theta)^2\big]\,d\theta \\
&\qquad + \int_\tau^T \big[(p-c_2)D(G_2^*(\tau,\theta)) - \frac{\kappa}{2}a_2^*(\tau,\theta)^2\big]\,d\theta + \sigma G_2^*(\tau,T)\Big)
\end{aligned}$$

$$+ \chi_{\tau \geq T} \Big( \int_t^T [(p - c_1) D(G_1^*(\theta)) - \frac{\kappa}{2} a_1^*(\theta)^2] \, dt + \sigma G_1^*(T) \Big) \, \Big| \, \tau > t \Big]$$

To write this in terms of value functions, we denote (as in Buratto et al., 2023) $V_2(s, t, G)$ for Stage 2 and $V(t, G)$ for Stage 1[2] and obtain

$$\xi_z(s, t) = V_2(s, t, G_2^*(s, t)), \qquad \lambda_z(t) = V(t, G_1^*(t)) \tag{7.19}$$

Having observed this, we can interpret the difference $[\xi_z(t, t) - \lambda_z(t)]$, which occurs frequently in the co-state system, as the "desirability" of the switch at time $t$. Let us denote it by $\lambda_\tau(t)$ (recall that $G_2^*(t, t) = G_1^*(t)$):

$$\begin{aligned} \lambda_\tau(t) :&= \xi_z(t, t) - \lambda_z(t) \\ &= V_2(t, t, G_1^*(t)) - V(t, G_1^*(t)) \end{aligned} \tag{7.20}$$

The reasoning behind this interpretation is that $\lambda_\tau(t)$ measures the expected gain in profit if the switch were to occur at time $t$, given that it has not occurred up to $t$. So, for example, $\lambda_\tau(t) < 0$ means that the profit that would be realized if the switch were to occur at time $t$ is lower than the expected profit, given that the system is still in Stage 1 at time $t$. This can be intuitively translated as "at time $t$, the switch is undesirable." Vice versa, if $\lambda_\tau(t) > 0$, the switch is desirable because the realized profit if $\tau = t$ is higher than the expected profit conditional on $\tau > t$ (i.e., the system is still in Stage 1 at time $t$).

In light of this, an undesirable (resp., desirable) switch has a backloading (resp., frontloading) effect on the Stage 1 goodwill's shadow value $\lambda_G$ (and therefore on $a_1$) that is proportional to the marginal hazard with respect to $G$ and the desirability of the switch. Intuitively, a farsighted planner will postpone (resp., advance) his advertising effort compared to a myopic planner, knowing that the switch will have a negative (resp., positive) effect on their optimal payoff. In this model, where $c_2 > c_1$, the switch turns out to be undesirable,

By comparing the ODEs for $\xi_G$ and $\lambda_G$, with the PDEs for the value functions $V_2$ and $V$, one can prove that

$$\xi_G(s, t) = \partial_G V_2(s, t, G_2^*(s, t)), \qquad \lambda_G(t) = \partial_G V(t, G_1^*(t)) \tag{7.21}$$

and therefore (recalling that $G_2^*(t, t) = G_1^*(t)$)

[2] In the cited paper, the Stage 1 problem which originates from plugging $V_2$ into the Stage 1 objective value (backward approach) is not solved, as it is a simple single-stage problem that the reader can solve with either dynamic programming or Pontryagin's maximum principle. With dynamic programming, one obtains a value function of the form $V(t, G, z) = z V^c(t, G)$. With the same abuse of notation as we employed for the co-state variables, we denote $V(t, G)$ the current value function $V^c(t, G)$.

$$\xi_G(t,t) - \lambda_G(t) = \partial_G\big[V_2(t,t,G_1^*(t)) - V(t,G_1^*(t))\big] \tag{7.22}$$

which is the expected marginal gain in profit from switching at time $t$ (given that the switch has not occurred yet) for a unit increment in the goodwill $G$ at time $t$. A negative expected marginal gain (i.e., $\xi_G(t,t) - \lambda_G(t) < 0$) means that, on average, at time $t$, a (slightly) higher goodwill than $G_1^*(t)$ would make switching at time $t$ (slightly) less convenient. For example, in the case that switching at time $t$ is undesirable for a given value of $G_1^*(t)$ (see previous paragraph), it may become even more undesirable if the goodwill were greater than $G_1^*(t)$.

From a different perspective, the same conclusion can be reached by observing that $\xi_G(t,t) - \lambda_G(t) = \kappa\big[a_2^*(t,t) - a_1^*(t)\big]$ If $\xi_G(t,t) - \lambda_G(t) < 0$ (resp., $>$), then the anticipation of the switch has a backloading (resp., frontloading) effect on $\lambda_G$ (and therefore on $a_1$). Intuitively, a non-myopic planner will postpone (resp., hasten) his advertising compared to a myopic planner, knowing that a higher goodwill would make the switch less (resp., more) convenient.

**Remark 7.2** It is worth highlighting how the special structure of this model simplifies the necessary conditions and, as a consequence, the solution of the switching time problem. Indeed, the adjoint equation for $\xi_G$ does not depend explicitly on $s$ because the problem is autonomous. Additionally, due to the linear state structure of the model, $G_2^*(s,t)$ (which would introduce an implicit, nontrivial dependence on $s$) does not factor into the equation either. Consequently, the dependence of the solution $\xi_G(s,t)$ on $s$ is trivial, as is that of the strategy $a_2^*(s,t)$.

In order to characterize the optimal advertising efforts, we need to solve the forward-backward system of the state and co-state dynamics constituted by (7.13) and (7.16). Observe that such a system is nonlinear, due, for example, to the presence of the multiplicative term between $\eta$ (depending on $G$) and $\lambda_G$; therefore, we resort to a numerical solution. In Sect. 7.5 we report some graphics with the optimal controls, states, co-states, and payoffs obtained with the numerical simulations.

The problem we have discussed so far relates to a planner who has all the information about the upcoming $\tau$ (hazard rate function and effects on the system) and has the ability to update his strategy upon the occurrence of $\tau$. In what follows, we will discuss planners with no information about $\tau$ or without the ability to update their strategy to the new regime after $\tau$. We will see how each of them has a specific functional objective, leading to different optimal strategies. This, of course, entails different expected payoffs, which, intuitively, are increasing with the planner's knowledge/ability.

## 7.4 Myopic Decision Makers

A planner is considered myopic if it does not take into account the possibility of a switch. Technically, myopic decision makers consider only the first equation of the system (7.13), and this definition is consistent with the game-theoretical one, by which a myopic player ignores the dynamics of a certain state variable; see, e.g., Taboubi and Zaccour (2002). Formally, their problem corresponds to (7.12) and (7.13) in which the hazard rate is identically 0; hence, the two auxiliary state variables $z_1$ and $z_2$ are constant and equal to 1 and 0, respectively.

Now, the question moves to the potential behavior of a myopic planner after the switch has occurred during the programming period. In light of these considerations, in the following definitions, we introduce two additional features that a myopic planner can have.

**Definition 7.1** A decision maker is *myopic with respect to a switching time and is unable to update their strategy* if, at the initial time, they compute their optimal advertising strategy by solving the single-stage problem (7.3) and cannot modify their strategy after the possible realization of the switching time.

In Definition 7.1, we are assuming that the myopic planner, once realized that $\tau$ has occurred and observed a sudden increase in production costs, may not be able to re-evaluate and update the advertising strategy according to the new regime. There may be several reasons for the impossibility to update the strategy to the abrupt event: They may have committed to a long-term advertising campaign or have contractual obligations with advertising agencies. Ultimately, the decision to continue to advertise a product at the same intensity, despite higher production costs, depends on several factors, including market conditions, competitive landscape, brand strategy, and long-term business goals.

**Theorem 7.2** *Let $\left(a^*(t), G^*(t)\right)$ be the optimal path in the optimal control problem (7.3) for decision makers who are myopic regarding switching time and are unable to update their strategy, then the optimal control function is*

$$a^*(t) = \frac{1}{\kappa}\Big(\frac{(p-c_1)\varepsilon\beta}{\delta}(1-e^{-\delta(T-t)}) + \sigma e^{-\delta(T-t)}\Big), \quad t \in [0, T] \tag{7.23}$$

***Proof*** Let us solve (7.3) using the necessary standard conditions (Grass et al., 2008, Th.3.4, p.109). Let us introduce the Hamiltonian function:

$$H(G, a, \lambda_G) = \Big[(p-c_1)\,\varepsilon(\alpha+\beta G) - \kappa a^2/2\Big] + \lambda_G(a - \delta G)$$

Maximizing with respect to $a \geq 0$ we obtain

$$a(t) = \left[\lambda_G(t)/\kappa\right]^+$$

with co-state equation satisfying

$$\begin{cases} \dot{\lambda}_G(t) = -(p-c_1)\varepsilon\beta + \delta\lambda_G(t) \\ \lambda_G(T) = \sigma \end{cases}$$

By a direct integration we get

$$\lambda_G(t) = \frac{(p - c_1)\varepsilon\beta}{\delta}(1 - e^{-\delta(T-t)}) + \sigma e^{-\delta(T-t)} > 0$$

Since $\lambda_G(t) > 0$ for all $t$, the optimal control turns out to be (7.23). □

Note that while for the non-myopic decision maker the optimal controls are characterized by numerically solving a system of ODEs, the optimal control of the myopic decision maker can be calculated explicitly.

Once the optimal solutions for both non-myopic and myopic planners are obtained, to evaluate the cost of myopia, we need to compare their expected profits. It is worth observing that even though the decision maker is myopic with respect to the switching time, such a random event can still occur, and therefore, the profit that we need to consider in the comparison is in any case an expected value. To be precise, we need to use the optimal control $a^*(t)$ to calculate its associated optimal state $G^*(t)$ and then determine the probability distribution of the switching time $\tau$. The differential equation governing the evolution of goodwill is linear, and therefore, it is possible to calculate the optimal state function explicitly. Recall that the optimal control for myopic decision makers who do not adapt to the new regime remains the same before and after the switching time. Furthermore, the dynamics does not change upon the switch, so the corresponding objective functional can be deduced by (7.7) where the denomination of the optimal control function $a^*(t)$ and its corresponding optimal state $G^*(t)$ do not change in the two stages.

$$\begin{aligned} J^*_{\text{Myopic}} = \mathbb{E}_{a^*(t)}\Big[ \mathbb{1}_{\{\tau<T\}} \Big\{ \int_0^\tau \big[(p - c_1)D(G^*(t)) - \frac{\kappa}{2}a^*(t)^2\big]\,dt \\ + \int_\tau^T \big[(p - c_2)D(G^*(t)) - \frac{\kappa}{2}a^*(t)^2\big]\,dt \Big\} \\ + \mathbb{1}_{\{\tau \geq T\}} \Big\{ \int_0^T \big[(p - c_1)D(G^*(t)) - \frac{\kappa}{2}a^*(t)^2\big]\,dt \Big\} \Big] + \sigma G^*(T) \end{aligned}$$

Denoting by $f^*_\tau(t)$ the density function of the random variable $\tau$, and by $F^*_\tau(t)$ its cumulative distribution function, we can explicitly calculate the expected value of the profit for the myopic decision maker who cannot adapt after the switch. Using the previous notation, we get

$$\begin{aligned} J^*_{\text{Myopic}} = \int_0^T f^*_\tau(s) \Big\{ \int_0^s \big[(p - c_1)D(G^*(t)) - \frac{\kappa}{2}a^*(t)^2\big]\,dt \\ + \int_s^T \big[(p - c_2)D(G^*(t)) - \frac{\kappa}{2}a^*(t)^2\big]\,dt \Big\}\,ds \\ + (1 - F^*_\tau(T)) \Big\{ \int_0^T \big[(p - c_1)D(G^*(t)) - \frac{\kappa}{2}a^*(t)^2\big]\,dt \Big\} \Big] + \sigma G^*(T) \end{aligned}$$

After integrating by parts the first line in the expression above we obtain the following:

$$J^*_{\text{Myopic}} = \int_0^T \mathbb{P}(\tau > t)\big[(p - c_1)D(G^*(t)) - \frac{\kappa}{2}a^*(t)^2\big]\,dt$$
$$+ \int_0^T \int_s^T f_\tau(s)\big[(p - c_2)D(G^*(t)) - \frac{\kappa}{2}a^*(t)^2\big]\,dt\,ds + \sigma G^*(T)$$

So far, we have assumed that the decision maker is unable to adjust its control in response to a change in production cost. Let us now provide a further definition that describes a different feature for a myopic decision maker who can update his control after the occurrence of the switch time.

**Definition 7.2** A decision maker is *myopic with respect to a switching time, but is able to update their strategy* if, at the initial time, he computes his optimal advertising strategy by solving the single-stage problem (7.3) and uses this strategy until the possible occurrence of the switching time. After the possible realization of the switching time, he updates his strategy by solving a new optimal control starting from the state level that is achieved at the realization of the switching time.

For this kind of decision maker, we need to provide two controls: one to be used before the realization of the switching time and the other afterward. It is useful to express these two controls in a way that depends on the random variable $\tau$.

**Theorem 7.3** *Let $a^\sharp(t)$ be the optimal strategy for a decision maker who is myopic with respect to a switching time, but is able to update his strategy, then the optimal control is*

$$a^\sharp(t) = \begin{cases} a^*(t) & t \in [0, \tau) \\ \frac{1}{\kappa}\left(\frac{(p-c_2)\varepsilon\beta}{\delta}(1 - e^{-\delta(T-t)}) + \sigma e^{-\delta(T-t)}\right) & t \in [\tau, T] \end{cases} \tag{7.24}$$

***Proof*** This results come straightforward from Theorem 7.2. A crucial point is in the form of the necessary conditions for the problem (7.3). The optimal control depends only on the co-state variable, which in turns is decoupled from the motion equation and can therefore be independently solved backward. □

Using the same notation and the same calculations described in this section, we can find the expected value for the decision maker who is myopic with respect to a switching time but who can update his strategy.

$$J^\sharp_{\text{Myopic}} = \int_0^T f^*_\tau(s)\Big\{\int_0^s \big[(p - c_1)D(G^*(t)) - \frac{\kappa}{2}a^*(t)^2\big]\,dt$$
$$+ \int_s^T \big[(p - c_2)D(G^\sharp(s,t)) - \frac{\kappa}{2}a^\sharp(t)^2\big]\,dt + \sigma G^\sharp_2(s, T)\Big\}\,ds$$

$$+ (1 - F_\tau^*(T))\Big\{ \int_0^T \big[(p - c_1)D(G^*(t)) - \frac{\kappa}{2} a^*(t)^2\big]\, dt + \sigma G^*(T)\Big\}\Big]$$

After integrating by parts the first line in the expression above, we obtain the following:

$$\begin{aligned} J^\sharp_{\text{Myopic}} &= \int_0^T \mathbb{P}(\tau > t)\big[(p - c_1)D(G^*(t)) - \frac{\kappa}{2} a^*(t)^2\big]\, dt + \mathbb{P}(\tau > T)\sigma G^*(T) \\ &+ \int_0^T \Big[\int_s^T f_\tau(s)\big[(p - c_2)D(G^\sharp(s,t)) - \frac{\kappa}{2} a^\sharp(t)^2\big]\, dt \\ &+ f_\tau(s)\sigma G^\sharp(s,T)\Big]\, ds \end{aligned}$$

In this section, we have characterized optimal solutions for a myopic decision maker in a closed form. However, to perform the comparison with respect to the non-myopic planner, numerical simulations will be necessary.

It is interesting to observe that in the single-stage optimal control problem solved by the myopic planners in Stage 1, their goodwill co-state functions satisfy the same Cauchy problem (7.16) of the non-myopic planner except for the value of epsilon which is $\varepsilon = 0$. The co-state equations for $\lambda_G(t)$ become

$$\begin{cases} \dot{\lambda}_G(t) = -(p - c_1)\pi + \delta\lambda_G(t) \\ \lambda_G(T) = \sigma \end{cases} \tag{7.25}$$

## 7.5 The Cost of Myopia

In this section, our objective is to draw a comparison between a myopic approach and a non-myopic one, addressing the research questions outlined in the Introduction. Specifically, we aim to quantify the cost of adopting a myopic perspective with respect to the switching time. With this attempt, we proceed numerically by assigning fixed values to some parameters and letting the marginal hazard with respect to the demand ($\varepsilon$) take values in $\{0.005, 0.01, 0.04\}$. The higher $\varepsilon$ the higher the hazard risk $\eta$ and therefore the more likely is the switch, which in our model corresponds to an increase in the production cost. The simulations have been produced with an algorithm based on Freiberger (2023). Figure 7.2 collects the parameters that are kept constant in all numerical simulations (Table 7.2).

**Table 7.2** Parameter values

| $p$ | $c_1$ | $c_2$ | $\alpha$ | $\beta$ | $\kappa$ | $\sigma$ | $\delta$ |
|---|---|---|---|---|---|---|---|
| 1 | 0.5 | 0.8 | 1 | 0.5 | 0.3 | 1 | 0.05 |

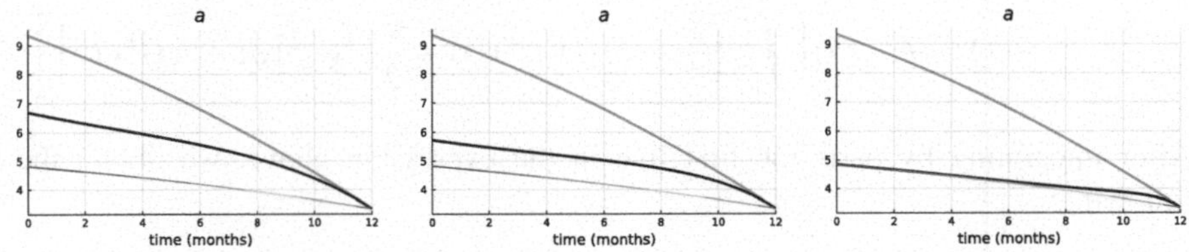


**Fig. 7.1** Optimal advertising efforts, $\varepsilon = 0.005$, $\varepsilon = 0.01$, $\varepsilon = 0.04$

In the following figures, the horizontal axes denote time (in months).

In the figures presented below, the myopic trajectories are represented by gray lines. The myopic planner who does not adapt will follow these trajectories even in the event of a regime shift. Conversely, the myopic planner who adapts will interrupt such strategies upon occurrence of the switching time. The non-myopic trajectories are depicted by black lines. Just like the myopic planner who adapts to the regime shift, the non-myopic planner will follow these lines throughout Stage 1, until the switching time occurs. Finally, the colored lines represent the Stage 2 behavior for different realizations of $\tau$. In the case of the myopic planner who adapts, these lines are slightly transparent, whereas for the non-myopic planner, they are presented in solid color. It is important to note that the Stage 2 advertising aligns in both scenarios. So, assuming that the switch occurs at a given $\overline{\tau}$, the optimal trajectory can be observed following the black line until the instant $\overline{\tau}$ and then "jumping" to the colored line from $\overline{\tau}$ on.

At a glance, we can observe that in both Figs. 7.1 and 7.2, the black lines are lower than the gray lines, regardless of $\varepsilon$. This is because a non-myopic planner knows that higher demand increases the likelihood of a switch, leading to increased costs. Consequently, the optimal advertising efforts of the non-myopic planner in Stage 1 turn are less intensive compared to those of the myopic planner. A similar pattern is observed in the goodwill trajectories in Fig. 7.2.

Moreover, in Fig. 7.1 it is worth noting that the myopic trajectories remain constant with changes in $\varepsilon$, while the non-myopic trajectories decrease as $\varepsilon$ increases. This indicates that the higher the marginal hazard with respect to the demand, the lower the advertising effort of the non-myopic planner. Again, the goodwill trajectories in Fig. 7.2 exhibit the same pattern.

In Fig. 7.2 each colored line starts at a given switch instant; therefore, the optimal goodwill trajectory can be seen following the black line from the initial time zero until the instant of the switch, while upon that instant, the associated colored line has to be considered. Each figure shows that the later the switch, the higher the optimal goodwill, while comparing the three figures it appears that the higher the marginal risk, the lower the optimal goodwill, as expected.

In Fig. 7.3 the objective values related to all the different types of behavior are plotted, with the further specific features the myopic planner can have, referring to Definitions 7.1 and 7.2: Gray =myopic+can update; Light gray =myopic+cannot

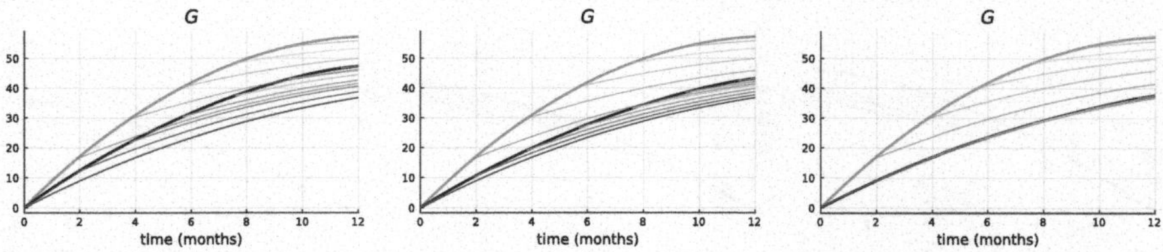


**Fig. 7.2** Optimal goodwill trajectories, $\varepsilon = 0.005$, $\varepsilon = 0.01$, $\varepsilon = 0.04$

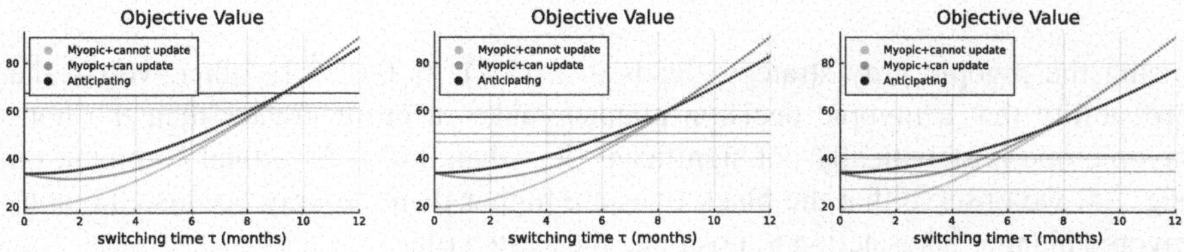


**Fig. 7.3** Profits; $\varepsilon = 0.005$ ($\tilde{\tau} \approx 9.5$), $\varepsilon = 0.01$ ($\tilde{\tau} \approx 8.5$), $\varepsilon = 0.04$ ($\tilde{\tau} \approx 7.5$)

update. Expected profits (represented by the solid constant lines) and realized profits (represented by the dotted lines) are shown as functions of the realization of the switch time. As before, the black color stands for non-myopic. Naturally, expected payoffs do not depend on the realization of $\tau$ and are therefore represented by constant lines.

The task of our analysis is to assess the cost of being myopic with respect to an abrupt switch. Indeed, we found that the expected profit for the non-myopic planner is greater than the ones of the myopic ones. Moreover, among myopic decision planners, the one who is able to update obtains higher profit because they can adapt to the new situation by reacting to the increase in production costs through a decrease in their advertising investment.

However, regarding realized payoffs, for sufficiently small realizations of $\tau$ non-myopic payoff lays over the myopic ones; nevertheless, it is a well-known result that there may be late realizations of $\tau$ where myopic planners achieve a higher payoff than their non-myopic counterparts. The dotted lines effectively intersect as $\tau$ approaches $T$, and, after such an intersection (let us call it $\tilde{\tau}$), the black dotted lines lay below the gray ones. In any case, this situation occurs with a very low probability, as clearly represented in Fig. 7.4, where the probability ($z_1$) of still being in Stage 1 is represented as a function of time. In all graphs, this probability decreases with $t$, consistent with its definition. Moreover, as the hazard risk increases, then the probability of arriving at late realizations of $\tau$ that provide a higher profit for the myopic decision planner is very small. For example, from the third graph of Fig. 7.3 ($\varepsilon = 0.04$) we can observe that the dark gray dotted line (myopic who can update) intersects the black one in $\tilde{\tau} \approx 7.1$ which in Fig. 7.4

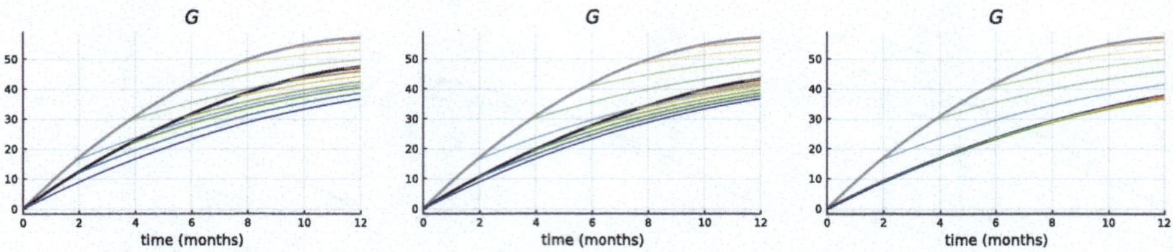


**Fig. 7.4** $z_1(t)$, $\varepsilon = 0.005$, $\varepsilon = 0.01$, $\varepsilon = 0.04$

(using the myopic gray strategy) leads to $z_1(7.1) \approx 0.02$.[3] In other words, the probability that a myopic decision planner makes a profit greater than the non-myopic one is statistically not significant (less than 0.05). As a final comment on Fig. 7.4, we observe that the black lines are higher than the gray because the non-myopic planner takes action to postpone the switch time.

## 7.6 Conclusions

This research offers a compelling analysis of the implications of myopia in decision making, particularly in the context of dynamic marketing problems. Our study centers on assessing the cost of a myopic approach when faced with potential abrupt changes in production costs, a scenario increasingly relevant in today's fast-paced and unpredictable market environments.

Through the formulation and solution of an optimal control problem with stochastic switching time, we were able to quantitatively compare the outcomes of a myopic planner with those of a planner who anticipates potential changes.

Referring to the first two research questions, our analysis shows how different decision-making approaches (myopic vs. non-myopic) affect advertising strategies and the resulting profits.

Second, we examine the impact of short-term thinking on expected profits. We evaluated the cost of myopia by assessing how a decrease in knowledge about crucial time-sensitive decisions affects profitability.

Our findings revealed a significant divergence in the profit outcomes between these two approaches. The myopic planner, constrained by a lack of foresight into possible system changes, invariably encountered a reduction in profit when the switch in production costs occurred. This reduction is directly attributable to the planner's inability to adjust strategies preemptively, highlighting the cost of myopia in decision making.

On the contrary, the non-myopic planner, equipped with the awareness of potential changes, demonstrated a more adaptable and resilient approach. This

[3] The value of $z_1$ evaluated in the $\tilde{\tau}$ corresponding to the intersection of the myopic who cannot update (light gray dotted line) is even smaller.

planner's ability to anticipate and plan for potential disruptions not only minimized losses but often led to more optimized use of resources, thereby maximizing profits. These results underscore the importance of strategic foresight in management and planning.

Our third research question: "With what probability does the myopic decision maker achieve a higher profit?" is intriguing, as it challenges the assumption that non-myopic (farsighted) planners always yield higher profits. The first answer can be found by observing that among myopic planners, there are those who can update their strategies after the switch and can indeed aim at achieving higher profits. However, in the cases of late switch realizations, myopic planners can occasionally outperform non-myopic ones. However, this scenario can occur with a low probability. In fact, our simulations prove that when the hazard rate is increased, the probability of myopic planners gaining higher profits in late switch realizations becomes statistically insignificant (less than 0.05 probability). This finding underscores that non-myopic planners, who proactively adapt to delay the switching time, generally yield better outcomes.

In conclusion, this paper highlights the tangible benefits of strategic anticipation and adaptability in dynamic decision-making contexts. It serves as a call to action for planners and managers to cultivate a forward-looking perspective, integrating predictive analytics and scenario planning into their strategic toolkit. By doing so, they can significantly reduce the risks associated with myopia and harness the full potential of their decision-making capabilities in an ever-evolving market landscape.

## References

Boukas, E.-K., Haurie, A., & Michel, P. (1990). An optimal control problem with a random stopping time. *Journal of Optimization Theory and Applications, 64*, 3, 471–480.

Brogan, W. L. (1968). Optimal control theory applied to systems described by partial differential equations. In *Advances in control systems* (Vol. 6, pp. 221–316). Elsevier.

Buratto, A., Muttoni, M., & Grosset, L. (2023). Two different solution techniques for an optimal control problem with a stochastic switching time. *WSEAS Transactions on Mathematics, 22*, 730–735.

Buratto, A., Muttoni, M., Wrzaczek, S., & Freiberger, M. (2022). Should the Covid-19 lockdown be relaxed or intensified in case a vaccine becomes available? *PLoS One, 17*, 9, e0273557.

Dawid, H., Keoula, M. Y., Kopel, M., & Kort, P. M. (2015). Product innovation incentives by an incumbent firm: A dynamic analysis. *Journal of Economic Behavior & Organization, 117*, 411–438.

Dockner, E., Jørgensen, S., Long, N. V., & Sorger, G. (2000). *Differential games in economics and management science*. Cambridge University Press.

Feichtinger, G., Hartl, R. F., & Sethi, S. P. (1994). Dynamic optimal control models in advertising: Recent developments. *Management Science, 40*(2), 195–226.

Feichtinger, G., Tragler, G., & Veliov, V. M. (2003). Optimality conditions for age-structured control systems. *Journal of Mathematical Analysis and Applications, 288*(1), 47–68.

Freiberger, M. (2023). Two-stage-optimal-control. Zenodo. https://doi.org/10.5281/zenodo.7772076

Grass, D., Caulkins, J. P., Feichtinger, G., Tragler, G., & Behrens, D. A. (2008). *Optimal control of nonlinear processes*. Springer.

Haunschmied, J. L., Kovacevic, R. M., Semmler, W., & Vladimir M., V. (2021). *Dynamic economic problems with regime switches*. Dynamic modeling and econometrics in economics and finance. Springer.

Huang, J., Leng, M., & Liang, L. (2012). Recent developments in dynamic advertising research. *European Journal of Operational Research, 220*(3), 591–609.

Kuhn, M., & Wrzaczek, S. (2021). Rationally risking addiction: A two-stage approach. In J. L. Haunschmied, R. M. Kovacevic, W. Semmler & V. M. Veliov, (Eds.), *Dynamic economic problems with regime switches*. Dynamic modeling and econometrics in economics and finance (pp. 85–110). Springer.

Nerlove, M., & Arrow, K. J. (1962). Optimal advertising policy under dynamic conditions. *Economica, 29*(114), 129–142.

Sethi, S. P. (1977). Dynamic optimal control models in advertising: A survey. *SIAM Review, 19*(4), 685–725.

Sorger, G. (1991). Maximum principle for control problems with uncertain horizon and variable discount rate. *Journal of Optimization Theory and Applications, 70*(3), 607–618.

Taboubi, S., & Zaccour, G. (2002). Impact of retailer's myopia on channel's strategies. In *Optimal control and differential games: Essays in honor of Steffen Jørgensen* (pp. 179–192). Springer.

Tsur, Y., & Zemel, A. (2017). Coping with multiple catastrophic threats. *Environmental and Resource Economics, 68*(1), 175–196.

Veliov, V. M. (2008). Optimal control of heterogeneous systems: basic theory. *Journal of Mathematical Analysis and Applications, 346*(1), 227–242.

Vidale, M. L., & Wolfe, H. B. (1957). An operations-research study of sales response to advertising. *Operations Research, 5*(3), 370–381.

Wang, P. (1964). Control of distributed parameter systems. *Advances in control systems, 1*, 75–172.

Wrzaczek, S., Kuhn, M., & Frankovic, I. (2020). Using age structure for a multi-stage optimal control model with random switching time. *Journal of Optimization Theory and Applications, 184*(3), 1065–1082.

# Chapter 8
# The Limits of Static Decision Rules in Supply Chain Games

**Fouad El Ouardighi, Suresh P. Sethi, and Christian Van Delft**

**Abstract** It is well known that static supply chain models ignore the future consequences of current actions. This shortsighted behavior can result either from the omission of the dynamics of important stock variables or from the use of extreme discounting. This chapter shows that the use of static supply chain models based on the omission of the evolution of important stock variables can lead to wrong decisions. To this end, we successively consider a series of simple issues representative of supply chain management. For each issue, two versions of a supply chain game are defined, one static and the other dynamic. For both versions, cooperative and non-cooperative scenarios are considered. For the static version, we do not use a naïve formulation but instead adopt an anticipatory perspective wherein the repetition of the static game over a given time horizon accounts for the update of the considered performance of the current period based on the previous period. Regarding the dynamic version, we use the framework of differential game theory. We then compare the decision rules and outcomes, respectively, inferred from the static and dynamic versions of the supply chain game considered. For each issue of interest, we show that the static decision rules provide distorted outcomes and misleading managerial prescriptions.

**Keywords** Supply chain management · Autonomous learning · Induced learning · Static decision rules · Optimal control · Differential games

---

F. El Ouardighi (✉)
ESSEC Business School, Cergy, France
e-mail: elouardighi@essec.edu

S. P. Sethi
The University of Texas at Dallas, Richardson, TX, USA

C. Van Delft
HEC, Jouy-en-Josas, France



F. El Ouardighi, G. Feichtinger (eds.), *The Unaffordable Price of Static Decision-making Models*, International Series in Operations Research & Management Science 365, https://doi.org/10.1007/978-3-031-88638-6_8

## 8.1 Introduction

Supply chain management poses a unique challenge due to the continuous interactions among different actors (e.g., El Ouardighi & Erickson, 2015; El Ouardighi & Schnaiderman, 2019; Han et al., 2024), i.e., suppliers *vs.* manufacturers and/or retailers, as well as among different management functions (e.g., El Ouardighi et al., 2008, 2013, 2016; Kennedy et al., 2021), i.e., operations management *vs.* marketing and/or finance, etc. These continuous interactions shape dynamical processes where any decision made at a specific instance reverberates through the overall system's future states.

Examples of supply chain management-related *issues* that require reliance on dynamic processes include marketing management, operational efficiency, inventory management, quality improvement, environmental sustainability, transactional conditions, and coordination among supply chain members. Though essential, these dynamical processes have been disregarded by a wide stream of the supply chain management literature for quite some time (e.g., Cachon & Fischer, 2000; Cachon, 2001; Corbett & Karmarkar, 2001; Chen, 2003).

Two widely cited articles may have reinforced this trend. The first paper surveyed the literature on supply chain coordination with contracts (Cachon, 2003). It does not refer to dynamic supply chain models; rather, the main objective is to minimize the instantaneous cost of non-cooperative behaviors of supply chain members. The second publication (Cachon & Netessine, 2004) provides an extensive review of the literature related to game theory in supply chain management. The emphasis is on static equilibrium strategies, and the reference to dynamic games is deferred to the very last pages of the paper, where the mathematical difficulty inherent to their analysis is mentioned repeatedly. To some extent, these reviews have, deliberately or not, contributed to forge the marked preference of the supply chain management literature for static models, thus contributing to the sidelining of seminal contributions by Jørgensen (1986), Eliashberg and Steinberg (1987), and Desai (1992).

This trend has certainly contributed to slowing the progress of dynamic analysis of supply chain management. In fact, although static supply chain games have remained prevalent in the literature (e.g., Nagarajan & Sošić, 2008, Zhao et al., 2010; Agi et al., 2021), an important stream of the supply chain literature promoting the use of dynamic games has emerged, almost against the grain of the dominant current. An early account of this research stream is provided in the studies by He et al. (2007) and Kogan and Tapiero (2007).

An intuitive reason for such unexpected success of dynamic supply chain games is because ignoring the role of dynamic processes in a supply chain (SC) leads one to infer misleading policies. An interesting illustration of such misleading policies is provided by Zaccour (2008), who showed that the implementation of a two-part tariff scheme in a static game model always generates an overconfident outcome in terms of the coordination of a marketing channel (see also Lambertini, 2014). Another intuitive reason for the development of dynamic SC games is that it allows one to more accurately articulate the continuous interactions between different management functions or activities within the SC, e.g., between operations and marketing activities (e.g., Desai, 1996) or between operations and environmental activities (e.g., El Ouardighi et al., 2016).

It is well known that static SC models ignore the future consequences of current actions. This shortsighted behavior, called myopia, can result either from the omission of the dynamics of important stock variables or from the use of extreme discounting. This chapter shows that the use of static SC models based on the omission of the evolution of important stock variables provides distorted outcomes and misleading managerial prescriptions. To this end, we successively consider a series of simple issues representative of supply chain management. For each issue, two deterministic versions of a SC game are defined, one static and the other dynamic. For both versions, we consider a cooperative and a non-cooperative scenario. Regarding the static version, we do not use a naïve formulation but instead adopt an anticipatory perspective wherein the repetition of the static game over a given time horizon accounts for the update of the considered performance of the current period based on the previous period. Regarding the dynamic version, we use the framework of differential game theory (Başar & Olsder, 1982; Dockner et al., 2000). We then compare the decision rules and outcomes, respectively, inferred from the static and dynamic versions of the SC game considered for the cooperative and non-cooperative scenarios, respectively. Our results show that static decision rules in an SC game:

- are not optimal,
- are not steady state compatible, and
- do not account for the information structure.

The chapter is structured as follows. In Sects. 8.2, 8.3, and 8.4, diverse SC dynamic game models, each associated with specific scenarios involving one supplier, one manufacturer, and/or one retailer, are thoroughly analyzed. We consider both shortsighted and farsighted decision rules. Within these examples, our analysis exposes the underlying mechanisms contributing to the discrepancy of static policies compared to their dynamic counterparts. Section 8.5 provides concluding remarks. In-depth technical developments are presented in the appendices.

## 8.2 Static Decision Rules in a Finite Time Horizon Supply Chain Game Are Not Optimal

The double marginalization effect in an SC under experience accumulation illustrates this point.

### 8.2.1 Model Formulation

We consider a build-to-order SC composed of a single manufacturer and a single supplier. The issue is: How can SC members' benefit be maximized in the presence of autonomous learning (i.e., experience effect) in the manufacturing process?

The supplier sells an intermediate good to the manufacturer, who uses it to produce a finished product. The manufacturer can reduce its unit operating cost over time thanks to its experience in production. The SC members can play either cooperatively or non-cooperatively. In the case of a non-cooperative game, they opt for a wholesale price contract (WPC), under which the supplier sets the transfer price. The WPC is supposed to be time-independent throughout the game horizon (El Ouardighi, 2014; El Ouardighi & Shnaiderman, 2019).

The state variable $X(t) \geq 0$, which represents the stock level of experience in production, evolves as:

$$\dot{X}(t) = S(t), \qquad X(0) = 0 \tag{8.1}$$

where $S(t) \geq 0$ is the consumer demand. We assume that consumer demand decreases linearly in the manufacturer's price $p_m(t) \geq 0$. That is,

$$S(t) = \alpha - \beta p_m(t) \tag{8.2}$$

where the subscript $m$ stands for the manufacturer. The parameter $a$ denotes the potential market and is supposed to be large, $\alpha \gg 0$. The parameter $\beta > 0$ is the marginal sensitivity of demand to consumer price. The experience is not subject to forgetting. The experience level reduces the manufacturer's production cost over time as follows (e.g., El Ouardighi et al., 2014):

$$C_m(t) = \omega_m - \theta X(t) \tag{8.3}$$

where $\omega_m > 0$ is the initial production cost and $\theta > 0$ denotes the marginal efficiency gain from autonomous learning.

We assume that $\omega_m$ is sufficiently large although lower than $\alpha/\beta$, both to prevent a negative operating cost and to serve as an incentive to develop experience. Finally, the supplier's transfer price is denoted by $p_s(t) = p_s$, $\forall 0 \leq t \leq T$, where $s$ stands for supplier, and the supplier's (constant) operating cost is $\omega_s \geq 0$. To get further insights, we normalize the supplier's cost to zero, i.e., $\omega_s = 0$.

Next, we define a profit function for each firm. We assume a fixed and finite planning horizon, $T < \infty$, and omit the discounting of future profits. Each firm aims to maximize its cumulative profits over the planning period.

Using (8.3), we write the manufacturer's instantaneous profit as:

$$[p_m(t) - (p_s(t) + \omega_m - \theta X(t))]\,(\alpha - \beta p_m(t)) \tag{8.4}$$

and the supplier's instantaneous gross profit as:

$$p_s(t)\,(\alpha - \beta p_m(t)) \tag{8.5}$$

Each SC member aims to maximize its profit in a dynamic non-cooperative game setting, that is:

$$\underset{p_m(t)\geq 0}{\text{Max}}\ \Pi_m^d = \int_0^T [p_m(t) - (p_s(t) + \omega_m - \theta X(t))]\, S(t)\mathrm{d}t \tag{8.6}$$

$$\underset{p_s(t)\geq 0}{\text{Max}}\ \Pi_s^d = \int_0^T p_s(t) S(t)\mathrm{d}t \tag{8.7}$$

subject to the experience dynamics in (8.1), where the superscript $d$ stands for dynamic.

Let us consider a static version of this non-cooperative game where the manufacturer sets a constant price over the whole planning horizon while anticipating the efficiency gain resulting from the production experience, as reflected in the cumulative sales from one period to the next. We assume that the efficiency gains are obtained with a one-period time lag. Given a finite time horizon $T < \infty$, the manufacturer's cumulative sales revenue is $p_m(\alpha - \beta p_m)T$, while the manufacturer's cumulative payment to the supplier and cumulative gross production cost are, respectively, $p_s(\alpha - \beta p_m)T$ and $\omega_m(\alpha - \beta p_m)T$. Regarding the instantaneous efficiency gains from autonomous learning, accounting for the time lag, they are defined as $X^s(t) = (\alpha - \beta p_m)t$, where $t \in [0, T - 1]$ and the superscript $s$ stands for static. At the end of the planning horizon, the cumulative efficiency gain from autonomous learning is $(\alpha - \beta p_m)\int_0^{T-1} t\mathrm{d}t = (\alpha - \beta p_m)\frac{(T-1)^2}{2}$. However, if we calculate the cumulative efficiency gains from autonomous learning as a sum of successive natural integers, that is, $(\alpha - \beta p_m)\sum_0^{T-1} t$, we get $(\alpha - \beta p_m)\frac{(T-1)T}{2}$, which is larger than $(\alpha - \beta p_m)\frac{(T-1)^2}{2}$. We choose this larger value for our analysis of the static game.

Therefore, the cumulative profit of the SC members in the static game is:

$$\underset{p_m\geq 0}{\text{Max}}\,\Pi_m^s = \left\{[p_m - (p_s + \omega_m)] + \theta\,(\alpha - \beta p)\,\frac{(T-1)}{2}\right\}(\alpha - \beta p_m)\,T \tag{8.8}$$

$$\underset{p_s\geq 0}{\text{Max}}\,\Pi_s^s = p_s\,(\alpha - \beta p_m)\,T \tag{8.9}$$

To determine a benchmark for the performance of the decentralized SC, we formulate the centralized problem, respectively, for the static and dynamic versions of the game. Because it is supposed to reflect the performance of a perfectly coordinated SC, the centralized problem requires that the SC members align their respective interests and jointly maximize the overall cumulative profits. For the dynamic setting, the centralized problem is:

$$\underset{p(t)\geq 0}{\text{Max}}\, \Pi_c^d = \int_0^T \left[p(t) - (\omega_m - \theta X(t))\right] \dot{X}(t)\mathrm{d}t \tag{8.10}$$

subject to (8.1), where the subscript $c$ stands for cooperative.

For the static setting, the centralized problem is:

$$\underset{p\geq 0}{\text{Max}}\Pi_c^s = \left[(p - \omega_m) + \theta\,(\alpha - \beta p)\,\frac{(T-1)}{2}\right](\alpha - \beta p)\,T \tag{8.11}$$

## 8.2.2 *Analysis*

### 8.2.2.1 The Cooperative Supply Chain

To ensure the solutions' feasibility and compare performances, we must have the same planning horizon for the four cases considered. This is why we choose the shorter of the dynamic cooperative setting time horizons ($T < \frac{1}{\beta\theta}$) and that of the dynamic non-cooperative setting ($T < 1 + \frac{3}{2\beta\theta}$). Thus, we set $T < \frac{1}{\beta\theta}$, for which we assume $\beta\theta \ll 1$.

**Lemma 1** The optimal static cooperative price is given by:

$$p_c^s = \frac{[1 - \beta\theta\,(T-1)]\,\alpha + \beta\omega_m}{\beta\,[2 - \beta\theta\,(T-1)]} > 0 \tag{8.12}$$

with the corresponding cooperative sales:

$$S_c^s = \frac{\alpha - \beta\omega_m}{2 - \beta\theta\,(T-1)} > 0 \tag{8.13}$$

and the maximized cooperative profit:

$$\Pi_c^s = \frac{(\alpha - \beta\omega_m)^2 T}{2\beta\,[2 - \beta\theta\,(T-1)]} > 0 \tag{8.14}$$

**Proof. A.1.1**

We get the following results by turning to the dynamic cooperative setting and skipping the time index for convenience.

**Lemma 2** The optimal farsighted cooperative price is given by:

$$p_c^d = \frac{(1 - \beta\theta T)\,\alpha + \beta\omega_m}{\beta\,(2 - \beta\theta T)} > 0 \tag{8.15}$$

with the corresponding cooperative sales:

$$S_c^d = \frac{\alpha - \beta \omega_m}{2 - \beta \theta T} > 0 \tag{8.16}$$

and the maximized cumulative cooperative profit:

$$\Pi_c^d = \frac{(\alpha - \beta \omega_m)^2 T}{2\beta (2 - \beta \theta T)} > 0 \tag{8.17}$$

**Proof. A.1.2**

Note that $p_c^d$ in (8.15) is time-independent, implying that the farsighted approach accounting for experience accumulation is compatible with a constant sales price. To ensure the non-negativity of the manufacturer's operating cost, i.e., $\omega_m - \theta X(t) \geq 0$, for $t \in [0, T]$, we require $\omega_m \geq \theta \frac{(\alpha - \beta \omega_m) T}{2 - \beta \theta T}$ at $t = T$. To avoid unnecessary technicalities, we assume a short enough planning horizon $T \leq \frac{2\omega_m}{\theta \alpha} < \frac{1}{\beta \theta} \iff \alpha > 2\beta \omega_m$. In this case, it is reasonable to use no discounting.

Comparing the farsighted cooperative pricing rule in (8.15) with the static cooperative one in (8.12), we get:

$$p_c^d - p_c^s = \frac{(1 - \beta \theta T) \alpha + \beta \omega_m}{\beta (2 - \beta \theta T)} - \frac{[1 - \beta \theta (T - 1)] \alpha + \beta \omega_m}{\beta [2 - \beta \theta (T - 1)]} < 0 \tag{8.18}$$

Thus, the time-independent farsighted pricing policy results in a larger consumer surplus than the static policy under SC cooperation. That is, the static pricing policy, on account of anticipating the impact of future experience on the profit function, is inefficient for the cooperative SC because it results in overpricing and, thus, underselling, as confirmed by the difference between (8.16) and (8.13).

In contrast, comparing (8.17) with the static cooperative cumulative profit in (8.14), we get:

$$\Pi_c^d - \Pi_c^s = \frac{(\alpha - \beta \omega_m)^2 T}{2\beta (2 - \beta \theta T)} - \frac{(\alpha - \beta \omega_m)^2 T}{2\beta [2 - \beta \theta (T - 1)]} > 0 \tag{8.19}$$

A farsighted policy does not require time-varying control, but is more profitable than the static policy under SC cooperation. Conversely, although it anticipates the impact of future experience on the pricing policy, the static decision rule is profit-ineffective for a cooperative SC.

#### 8.2.2.2 Non-cooperative Supply Chain

We now consider the case where the SC members agree on a wholesale price contract (WPC). Because the supplier sets the wholesale price, it is the Stackelberg leader.

We first handle the static non-cooperative setting.

**Lemma 3** The manufacturer's static non-cooperative price is given by:

$$p_m^s = \frac{[3 - 2\beta\theta (T - 1)]\,\alpha + \beta\omega_m}{2\beta\,[2 - \beta\theta (T - 1)]} > 0 \tag{8.20}$$

while the supplier's non-cooperative wholesale price is:

$$p_s^s = \frac{\alpha - \beta\omega_m}{2\beta} > 0 \tag{8.21}$$

The corresponding non-cooperative sales are:

$$S_m^s = \frac{\alpha - \beta\omega_m}{2\,[2 - \beta\theta (T - 1)]} > 0 \tag{8.22}$$

and the maximized non-cooperative static profits are, respectively,

$$\Pi_m^s = \frac{(\alpha - \beta\omega_m)^2 T}{8\beta\,[2 - \beta\theta (T - 1)]} > 0 \tag{8.23}$$

$$\Pi_s^s = \frac{(\alpha - \beta\omega_m)^2 T}{4\beta\,[2 - \beta\theta (T - 1)]} > 0. \tag{8.24}$$

**Proof. A.1.3**
Notably, the supplier gets twice as much profit as the manufacturer. We can now assess each SC member's incentive to cooperate in the static setting.

**Lemma 4** The manufacturer's and supplier's static cooperative profits are, respectively,

$$\Pi_m^{sc} = \frac{3(\alpha - \beta\omega_m)^2 T}{16\beta\,[2 - \beta\theta (T - 1)]} \tag{8.25}$$

$$\Pi_s^{sc} = \frac{5(\alpha - \beta\omega_m)^2 T}{16\beta\,[2 - \beta\theta (T - 1)]} \tag{8.26}$$

where the superscript $c$ denotes cooperative.

**Proof. A.1.4**
Comparing (8.25) and (8.26), we observe that the supplier has a greater incentive for SC cooperation than does the manufacturer.

We now turn to the dynamic non-cooperative setting. In practice, a contract is often agreed on from the outset of the game to make the SC members' contractual relations as predictable as possible throughout a finite planning horizon. Because the autonomous learning is supposed to reduce the manufacturer's operating cost only, it is plausible to assume that the supplier ignores its intertemporal evolution, if not its existence, and thus behaves myopically with respect to the manufacturing learning effect. In this setup, the supplier sets a time-independent transfer price $p_s(t) = p_s$, $\forall 0 \leq t \leq T$.

Given that the supplier acts as a Stackelberg leader, a two-stage game is formulated in which the supplier chooses the transfer price at the first stage to maximize its profit. The manufacturer can then either accept the optimally set transfer price or not. In this setup, the game's second stage is played dynamically, and the two-stage game is solved backwardly (e.g., El Ouardighi & Shnaiderman, 2019). We restrict our attention to subgame perfect non-cooperative equilibrium and thus use dynamic programming to determine the manufacturer's non-cooperative strategy.

**Lemma 5** Under a time-independent wholesale price contract, the farsighted non-cooperative manufacturer's equilibrium sales price is given by:

$$p_m^d = \frac{(3 - 2\beta\theta T)\,\alpha + \beta\omega_m}{2\beta\,(2 - \beta\theta T)} > 0 \tag{8.27}$$

and the supplier's equilibrium wholesale price by:

$$p_s^d = \frac{\alpha - \beta\omega_m}{2\beta} > 0 \tag{8.28}$$

The corresponding non-cooperative sales are:

$$S_m^d = \frac{\alpha - \beta\omega_m}{2\,(2 - \beta\theta T)} > 0 \tag{8.29}$$

and the maximized non-cooperative profits are, respectively,

$$\Pi_m^d = \frac{(\alpha - \beta\omega_m)^2 T}{8\beta\,(2 - \beta\theta T)} > 0 \tag{8.30}$$

$$\Pi_s^d = \frac{(\alpha - \beta\omega_m)^2 T}{4\beta\,(2 - \beta\theta T)} > 0 \tag{8.31}$$

**Proof. A.1.5**
Note that the manufacturer's farsighted non-cooperative policy also results in a time-independent pricing control and thus constant sales over time. As shown in (A.2.2), the manufacturer's experience increases linearly. Finally, the supplier's profit is twice that of the manufacturer.

**Lemma 6** The manufacturer's and supplier's cooperative profits under a time-independent wholesale price contract are, respectively,

$$\Pi_m^{dc} = \frac{3(\alpha - \beta\omega_m)^2 T}{16\beta\,(2 - \beta\theta T)} \tag{8.32}$$

$$\Pi_s^{dc} = \frac{5(\alpha - \beta\omega_m)^2 T}{16\beta\,(2 - \beta\theta T)} \tag{8.33}$$

where the superscript $c$ denotes cooperative.

**Proof. A.1.6**

#### 8.2.2.3 Comparisons

Using (8.18) and (8.19), we conclude:

**Proposition 1** The social welfare gained from a centralized SC is more significant in the dynamic than in the static setting.

This result is driven by overpricing, as shown in (8.18), which weakens autonomous learning and lowers the benefits, as reported in (8.19). Consequently, the static decision rule cannot lead to optimal social welfare in SC cooperation. It will produce misleading managerial prescriptions for the best way to exploit autonomous learning.

Comparing (8.21) and (8.28), on the one hand, and (8.20) and (8.27), on the other hand, we get:

$$p_s^s = p_s^d \tag{8.34}$$

$$p_m^d < p_m^s \tag{8.35}$$

That is, the supplier's transfer price is the same in the static and dynamic settings. In addition, the farsighted non-cooperative manufacturer's price is lower than the static one. It implies a greater consumer surplus and sales in the dynamic non-cooperative case than in the static one.

Be it a static or dynamic game, it is not clear whether the supplier's transfer price is greater or lower than the cooperative sales price.

Comparing (8.27) and (8.15), on the one hand, and (8.20) and (8.12), on the other hand, we obtain:

$$p_m^d - p_c^d = \frac{\alpha - \beta\omega_m}{2\beta\,(2 - \beta\theta T)} > p_m^s - p_c^s = \frac{\alpha - \beta\omega_m}{2\beta\,[2 - \beta\theta\,(T - 2)]} \tag{8.36}$$

Despite an equal transfer price in the dynamic and the static setting, the difference between the manufacturer's and the cooperative price is greater in a dynamic setting than in a static setting.

Comparing (8.100), (8.107), (8.109), and (8.124) from Appendix 1 gives:

$$X_m^s(T) < X_m^d(T) < X_c^s(T) < X_c^d(T) \tag{8.37}$$

Therefore, we conclude:

**Proposition 2** The double marginalization effect is more potent in a dynamic than in a static SC. Nevertheless, autonomous learning is more effective under dynamic than static conditions.

The results in Proposition 2 are illustrated in Fig. 8.1.

Comparing (8.23) with (8.30), on the one hand, and (8.24) with (8.31), on the other hand, we get the following ranking:

$$\Pi_m^d > \Pi_m^s \tag{8.38}$$

$$\Pi_s^d > \Pi_s^s \tag{8.39}$$

Both SC members get a greater non-cooperative profit under a dynamic than in a static setting.

Therefore, using (8.35), (8.38), and (8.39), we conclude:

**Proposition 3** The social welfare drawn from a non-cooperative SC is greater in the dynamic than in the static setting.

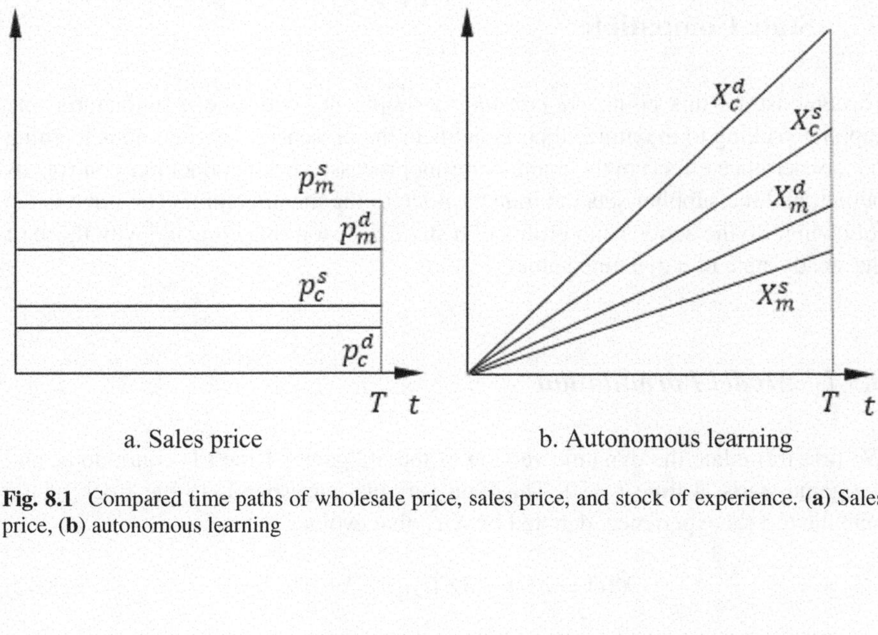


**Fig. 8.1** Compared time paths of wholesale price, sales price, and stock of experience. (**a**) Sales price, (**b**) autonomous learning

Finally, in a decentralized SC, static decision-making models also give rise to misleading managerial prescriptions on how to exploit the best autonomous learning by doing.

Comparing (8.25) with (8.32), on the one hand, and (8.26) with (8.33), on the other hand, we get:

$$\Pi_m^{dc} > \Pi_m^{sc} \tag{8.40}$$

$$\Pi_s^{dc} > \Pi_s^{sc} \tag{8.41}$$

In (8.40) and (8.41), the cooperative profits are greater in a dynamic than in a static setting.

Hence the following proposition:

**Proposition 4** The incentive for SC cooperation is greater for both players in a dynamic than in a static setting.

In other words, the static decision-making model in an SC underestimates the benefit that could be drawn from cooperation by the SC members, thus distorting their incentive to cooperate. The reason for this is the undervaluation of the double marginalization effect in a static setting, which reduces the SC members' motivation to mitigate the related inefficiency through cooperation.

## 8.3 Static Decision Rules in a Supply Chain Are Not Steady State Compatible

To demonstrate this point, we consider a simple SC with one manufacturer and supplier seeking to maximize their benefits in the presence of autonomous learning (i.e., experience effect) in the manufacturing process. The manufacturer controls its output, and the supplier sets the transfer price to the manufacturer. The issue is the following: do the actions and profits of a static decision rule coincide with those of the steady state of a dynamic game?

### *8.3.1 Model Formulation*

We first formulate the dynamic version of the SC game. Time $t$ is continuous, and the game starts at time $t = 0$. The state variable corresponds to the stock of the manufacturer's experience, denoted by $X(t)$, that evolves as:

$$\dot{X}(t) = q(t) - \delta X(t), \qquad X(0) = 0 \tag{8.42}$$

where $q(t) \geq 0$ is the manufacturer's output and $\delta > 0$ is a forgetting parameter.

Let us now define a payoff function for each firm. The experience level is supposed to reduce the manufacturer's initial operating cost linearly, $\omega_m > 0$, that is (e.g., El Ouardighi et al., 2014):

$$C_m(t) = \omega_m - \theta X(t) \tag{8.43}$$

where $\theta > 0$ denotes the marginal efficiency gain from autonomous learning, and the subscript $m$ stands for the manufacturer. We assume that the manufacturer's initial operating cost is sufficiently large, $\omega_m \gg 0$ to both prevent a negative operating cost and to serve as an incentive to benefit from experience.

The SC members agree to a wholesale price contract where the transfer price, $p_s(t) \geq 0$, the subscript $s$ denoting the supplier, is paid by the manufacturer to the supplier for each unit of input purchased, for an instantaneous volume of inputs purchased equivalent to the manufacturer's instantaneous output, $q(t)$. Here also, the supplier in the dynamic game is supposed to ignore the intertemporal evolution of autonomous learning, that is, to behave myopically with respect to the manufacturer's experience effect. Therefore, the supplier sets the transfer price once and for all at the initial period of the game, i.e., $p_s(t) = p_s$, for $t \geq 0$. Finally, the supplier's (constant) operating cost is $\omega_s \geq 0$. To get further insights, it is normalized to zero, i.e., $\omega_s = 0$.

Let the sales price be $p(t) = a - q(t)$, where the parameter $a > 0$ denotes the maximum potential price. Using (8.43), the manufacturer's instantaneous profit is:

$$[a - q(t) - (p_s + \omega_m - \theta X(t))]\, q(t)$$

and the supplier's instantaneous gross profit is:

$$p_s q(t)$$

To characterize the long-run equilibrium, i.e., the steady state, if it exists, of the SC, the planning horizon of the dynamic version of the game is supposed to be infinite, that is, $t \in [0, \infty[$. Assuming that both firms employ a symmetric discounting rate, $r > 0$, the maximization of the cumulative profit of the SC members in the dynamic game problem is then given as:

$$\underset{q(t)\geq 0}{\text{Max}}\ \Pi_m^d = \int_0^{\infty} e^{-rt}\, [a - q(t) - (p_s + \omega_m - \theta X(t))]\, q(t) \mathrm{d}t \tag{8.44}$$

$$\underset{p_s(t)\geq 0}{\text{Max}}\ \Pi_s^d = \int_0^{\infty} e^{-rt}\, p_s q(t) \mathrm{d}t \tag{8.45}$$

under the constraint (8.42), where the superscript $d$ stands for dynamic.

In contrast with the dynamic setting above, we assume that the static decision rules are implemented at a given time period $t$ where $1 < t \leq T < \infty$. At $t = T$, the maximum instantaneous efficiency gains are drawn from autonomous learning, that is, as $X(t) = (T - 1)q$. For the sake of simplicity, discounting can be omitted in the static version of the model.

The respective instantaneous profits of the SC members in the static game setting are:

$$\underset{q \geq 0}{\text{Max}} \pi_m^s = [a - q - p_s - \omega_m + \theta q \, (t - 1)] \, q \tag{8.46}$$

$$\underset{p_s \geq 0}{\text{Max}} \pi_s^s = p_s q \tag{8.47}$$

where the superscript $s$ stands for static.

Finally, the centralized decision-making problem for the dynamic setting writes:

$$\underset{q(t) \geq 0}{\text{Max}} \; \Pi_c^d = \int_0^\infty e^{-rt} \, [a - q(t) - (\omega_m - \theta X(t))] \, q(t) \text{d}t \tag{8.48}$$

under the constraint (8.42), where the subscript $c$ stands for cooperative.

For the static setting, the centralized instantaneous profit is:

$$\underset{q \geq 0}{\text{Max}} \pi_c^s = [a - q - \omega_m + \theta q \, (t - 1)] \, q \tag{8.49}$$

## 8.3.2 *Analysis*

### 8.3.2.1 Cooperative Supply Chain

**Lemma 7** For $T < 1 + \frac{1}{\theta}$, the optimal static cooperative production and experience level are given by:

$$q_c^s = \frac{a - \omega_m}{2 \, [1 - \theta \, (T - 1)]} \tag{8.50}$$

$$X_c^s = \frac{(a - \omega_m) \, (T - 1)}{2 \, [1 - \theta \, (T - 1)]} \tag{8.51}$$

with the corresponding cooperative profit:

$$\pi_c^s = \frac{(a - \omega_m)^2}{4 \, [1 - \theta \, (T - 1)]} \tag{8.52}$$

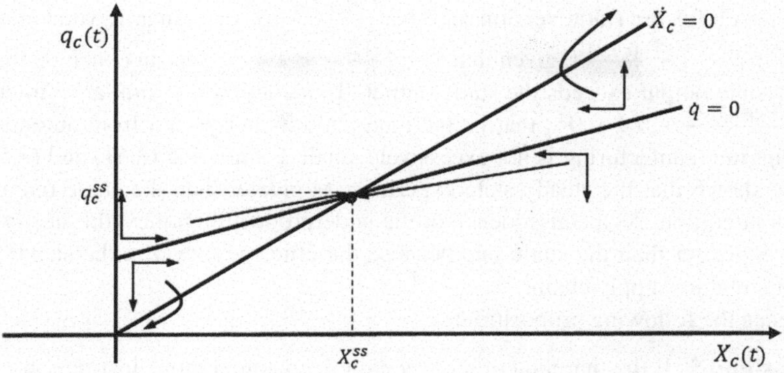


**Fig. 8.2** Phase-portrait diagram in the state-control space

**Proof. A 2.1**
We get the following results by turning to the dynamic cooperative setting and skipping the time index for convenience.

**Lemma 8** If the marginal benefit from the experience accumulated is sufficiently low, i.e., $\theta < \frac{2\delta(r+\delta)}{r+2\delta}$, there exists a locally stable steady state given by the following cooperative production rate, experience level, and profit:

$$q_c^{ss} = \frac{\delta\,(r+\delta)\,(a-\omega_m)}{2\delta\,(r+\delta)-\theta\,(r+2\delta)} \tag{8.53}$$

$$X_c^{ss} = \frac{(r+\delta)\,(a-\omega_m)}{2\delta\,(r+\delta)-\theta\,(r+2\delta)} \tag{8.54}$$

$$\pi_c^{ss} = \frac{\delta^2\,(r+\delta)\,(r+\delta-\theta)\,(a-\omega_m)^2}{[2\delta\,(r+\delta)-\theta\,(r+2\delta)]^2} \tag{8.55}$$

where the superscript *ss* stands for steady state. Otherwise, there is no feasible steady state.

**Proof. A.2.2**
The stability of convergence is contingent upon the magnitude of the marginal benefit from the experience effect, which is more in line with mature than emerging industries. The phase diagram in Fig. 8.2 depicts the qualitative properties of the solution obtained in the state-costate space. Starting with a relatively low (high) initial experience level, i.e., $X_0 < X_c^{ss}$ $\left(X_0 > X_c^{ss}\right)$, the optimal policy consists of setting an initially low (high) and increasing (decreasing) production rate to converge to the locally stable steady state.

Note that the static decision rules do not involve any condition for convergence toward the long-run cooperative state. By comparing (8.50) with (8.53), provided

$T < 1 + \frac{1}{\theta}$ for the static version and $\theta < \frac{2\delta(r+\delta)}{r+2\delta}$ for the dynamic version of the game, if $T < 1 + \frac{(r+2\delta)}{2\delta(r+\delta)}$, given that $1 + \frac{(r+2\delta)}{2\delta(r+\delta)} < 1 + \frac{1}{\theta}$, we can conclude that the steady state output exceeds the static output. This condition is similar to imposing $\theta \approx \frac{2\delta(r+\delta)}{r+2\delta} - \varepsilon$, $\varepsilon \to 0^+$, that is, the marginal efficiency gain from autonomous learning in manufacturing is not excessively small. Comparing (8.51) and (8.54), it can be shown that the steady state experience is greater than the static one under similar condition. As a consequence of the underproduction policy, the steady state profit is greater than the static one because the efficiency level at the steady state reflect a mature supply chain.

Hence the following proposition:

**Proposition 5** If the marginal efficiency gain from autonomous learning in manufacturing is not excessively small, static decision rules underestimate the long-run benefit from cooperative supply chain management.

#### 8.3.2.2 Non-cooperative Supply Chain

We now consider the case where the SC members play non-cooperatively.

We start with the static setting.

**Lemma 9** For $T < 1 + \frac{1}{\theta}$, the optimal static non-cooperative transfer price and output are given by:

$$p_s^s = \frac{a - \omega_m}{2} \tag{8.56}$$

$$q_{nc}^s = \frac{a - \omega_m}{4\left[1 - \theta\left(T - 1\right)\right]} \tag{8.57}$$

with the corresponding SC members' profit:

$$\pi_s^s = \frac{(a - \omega_m)^2}{8\left[1 - \theta\left(T - 1\right)\right]} \tag{8.58}$$

$$\pi_m^s = \frac{(a - \omega_m)^2}{16\left[1 - \theta\left(T - 1\right)\right]} \tag{8.59}$$

**Proof. A.2.3**

Considering the dynamic version of the decentralized game model, because of a myopic supplier, the SC members agree upon a WPC with a constant transfer price. Therefore, as the Stackelberg leader, the supplier initially sets the transfer price. Then, the manufacturer decides its production strategy over time. We use the Bellman principle to derive the manufacturer's (linear) feedback equilibrium strategy.

**Lemma 10** Under $\theta < \frac{2\delta(r+\delta)}{r+2\delta}$, the non-cooperative dynamic game has one globally asymptotically stable steady state, that is given by the following equilibrium transfer price, production rate, experience level, and steady state profits, that is:

$$p_s^{ss} = \frac{a - \omega_m}{2} \tag{8.60}$$

$$q_s^{ss} = \frac{\delta (r + \delta) (a - \omega_m)}{4\delta (r + \delta) - 2\theta (r + 2\delta)} \tag{8.61}$$

$$X_{nc}^{ss} = \frac{(r + \delta) (a - \omega_m)}{4\delta (r + \delta) - 2\theta (r + 2\delta)} \tag{8.62}$$

$$\pi_s^{ss} = \frac{\delta (r + \delta) (a - \omega_m)^2}{4 [2\delta (r + \delta) - \theta (r + 2\delta)]} \tag{8.63}$$

$$\pi_m^{ss} = \frac{\delta^2 (r + \delta) (r + \delta - \theta) (a - \omega_m)^2}{4[2\delta (r + \delta) - \theta (r + 2\delta)]^2} \tag{8.64}$$

**Proof. A.2.4**
Note that, in (8.64), the condition for a positive manufacturer's steady state profit, that is, $\theta < r + \delta$, is fulfilled under $\theta < \frac{2\delta(r+\delta)}{r+2\delta}$. Though (8.56) and (8.60) are equivalent, i.e., the supplier's transfer price is the same for both myopic and farsighted manufacturer, (8.57) and (8.61) are different, that is, the myopic and steady state production rates are different. Here also, if $\theta \approx \frac{2\delta(r+\delta)}{r+2\delta} - \varepsilon, \varepsilon \to 0^+$, that is, the marginal efficiency gain from autonomous learning in manufacturing is not excessively small, the myopic decision rule involves underproduction policy. The reason for this lies in the fact that the static decision rules disregard the dynamic nature of the strategic interactions among the SC members throughout the game. This impact results in greater efficiency gains and thus, greater steady state profits than in the static game.

**Proposition 6** If the marginal efficiency gain from autonomous learning in manufacturing is not excessively small, static decision rules underestimate the long-run efficiency gains of non-cooperative supply chain management.

We conclude that static decision rules cannot serve to accurately approximate a steady state equilibrium in an SC, because they lead to a distortion of the long-run social welfare, regardless of the players' mode of play.

## 8.4 Static Decision Rules in a Supply Chain Do Not Account for Information Structure

To provide evidence regarding this point, we again consider the case of a one manufacturer-one supplier SC. The issue is as follows: How to maximize the SC members' individual benefits derived from the effect of induced learning on the manufacturer's production cost? Here, the difference with the problem introduced in Sect. 8.2 is that efficiency is gained not through production experience accumulation (autonomous learning) but rather with quality improvement and/or R&D efforts, i.e., induced learning (e.g., Kogan & El Ouardighi, 2019).

### 8.4.1 Model Formulation

Let us consider the case of a simple SC with one manufacturer and one supplier. The SC members both invest in a cost-reducing R&D activity to decrease the manufacturer's operating cost. The stock of R&D, $X(t) \geq 0$, evolves according to:

$$\dot{X}(t) = u(t) + v(t), \qquad X(0) = X_0 \geq 0 \tag{8.65}$$

where $u(t)$, $v(t) \geq 0$, respectively, denote the supplier's and the manufacturer's cost-reducing R&D efforts. The SC members agree on a wholesale price contract with a time-independent wholesale price (e.g., El Ouardighi & Shnaiderman, 2019). Assuming that the final good is produced with a fixed-coefficient technology, i.e., one unit of the input for one unit of the final product, the consumer price, $p(t) \geq 0$, is defined as the sum of the constant supplier's transfer price, $p_s \geq 0$, and the manufacturer's current operating cost, $C_m(t) \geq 0$, and profit margin, $\pi_m(t) \geq 0$, that is:

$$p(t) = p_s + C_m(t) + \pi_m(t)$$

The supplier's profit margin, $\pi_s(t) \geq 0$, is given as the difference between the supplier's transfer price and (constant) operating cost, $\omega_s \geq 0$, that is:

$$\pi_s \equiv p_s - \omega_s \tag{8.66}$$

To get further insights, we normalize the supplier's cost to zero, i.e., $\omega_s = 0$, so that the supplier's transfer price is equal to its unit profit margin, that is, $\pi_s \equiv p_s$.

In addition, the R&D stock is supposed to reduce the manufacturer's initial operating cost, $\omega_m \geq 0$, linearly so that:

$$C_m(t) = \omega_m - \theta X(t) \tag{8.67}$$

where $\theta > 0$ denotes the marginal efficiency gain from induced learning.

Using the previous assumptions, we rewrite the consumer price as follows:

$$p(t) = \pi_s + \omega_m - \theta X(t) + \pi_m(t) \tag{8.68}$$

The consumer demand is supposed to be a linear negative function of price, that is:

$$S(t) = \alpha - \beta\,(\pi_s + \omega_m - \theta X(t) + \pi_m(t)) \tag{8.69}$$

Next, we define a profit function for each firm. We assume a fixed and finite planning horizon, $T < \infty$, and omit the discounting of future profits. Each firm aims to maximize its cumulative profits over the planning period.

Using (8.68)–(8.69), we write the manufacturer's gross profit as:

$$(p(t) - p_s + C_m(t))\, S(t) \equiv \pi_m(t)\,[\alpha - \beta\,(\pi_s + \omega_m - \theta X(t) + \pi_m(t))]$$

and using (8.67)–(8.69), the supplier's gross profit is:

$$p_s S(t) \equiv \pi_s\,[\alpha - \beta\,(\pi_s + \omega_m - \theta X(t) + \pi_m(t))]$$

Assuming that advertising investments are subject to diminishing returns, we suppose that the cost of effort in goodwill accumulation is an increasing quadratic function for both firms, that is, $e_s u(t)^2/2$ and $e_m v(t)^2/2$, with $e_s$, $e_m > 0$. Assuming that the SC members have comparable efficiency and/or access to funds to finance their respective R&D efforts and to keep the problem tractable, we let $e_s = e_m = 1$.

The differential game problem of the SC members is then rewritten as:

$$\underset{\pi_s, u_s(t)}{\text{Max}}\; J_s = \int_0^T \left\{ \pi_s\,[\alpha - \beta\,(\pi_s + \omega_m - \theta X(t) + \pi_m(t))] - \frac{u(t)^2}{2} \right\} \mathrm{d}t \tag{8.70}$$

$$\underset{\pi_m(t), u_m(t)}{\text{Max}}\; J_m = \int_0^T \left\{ \pi_m(t)\,[\alpha - \beta\,(\pi_s + \omega_m - \theta X(t) + \pi_m(t))] - \frac{v(t)^2}{2} \right\} \mathrm{d}t \tag{8.71}$$

Considering the static version of the above SC game, as for the case of autonomous learning, the manufacturer anticipates the efficiency gains drawn from the induced learning engendered by the SC members' R&D efforts at each time period. Here, we also assume that the efficiency gains are obtained with a one-period time lag. Given a time horizon $T < \infty$, the cumulative efficiency gains from induced learning are defined as $X^s(t) = (u + v)t$, where $t \in [0, T - 1]$ and $s$ stands for static. At the end of the planning horizon, the cumulative efficiency gains from induced learning are given by: $(u + v)\int_0^{T-1} t\,\mathrm{d}t = (u + v)\,\frac{(T-1)^2}{2}$. However, if we calculate the cumulative efficiency gains from induced learning as a sum of successive natural integers, that is, $(u+v)\sum_0^{T-1} t$, we get $(u+v)\frac{(T-1)T}{2}$, which is an upper value for $(u+v)\frac{(T-1)^2}{2}$. Here, we also choose this upper value for our computations of the static game.

Therefore, the cumulative profit of the SC members in the static setting is written as:

$$\underset{\pi_s, u_s \geq 0}{\text{Max}}\ \Pi_s^s = \pi_s \left[ \alpha - \beta \left( \pi_s + \omega_m - \theta\,(u+v)\,\frac{(T-1)}{2} + \pi_m \right) - \frac{u^2}{2} \right] T \tag{8.72}$$

$$\underset{\pi_m, u_m \geq 0}{\text{Max}}\ \Pi_m^s = \pi_m \left[ \alpha - \beta \left( \pi_s + \omega_m - \theta\,(u+v)\,\frac{(T-1)}{2} + \pi_m - \frac{v^2}{2} \right) \right] T \tag{8.73}$$

where the superscript $s$ stands for static.

Both the static and dynamic versions of the game involve simultaneous moves. One main difference between the two versions of the game lies with the fact that, in the dynamic model, the supplier acts as a Stackelberg leader at the first stage of the game by choosing the transfer price to maximize its individual profit. If the manufacturer accepts the optimal WPC, the second stage of the game is then played *à la Nash*. Another essential difference lies with the underlying information structure. That is, the resolution of the static game proceeds from a unique decision pattern. In contrast, the dynamic version of the game may reflect a wide spectrum of decision patterns that emerge from the alternative decision rules that are available to the SC participants, that is, open-loop Nash equilibrium (OLNE), closed-loop Nash equilibrium (CLNE), and feedback Nash equilibrium (FBNE). Depending on the decision rule chosen by each firm, the overall performance of the SC can vary significantly. The three decision rules differ in terms of how they adjust for changing values of state variables (Başar & Olsder, 1999; Dockner et al., 2000; Long, 2010). In an OLNE, firms select their strategies at the beginning of the game and commit to them thereafter. That is, the strategic interaction between the players is circumscribed to the initial time period. Given state variables' initial values, OLNE decisions depend only on time. In a CLNE, controls are functions of initial as well as current values of the state variables. In a FBNE, controls depend only on current values of the state variables. Since the other players' strategies affect the current value of state variables, taking into account the current state vector in FBNE and, in general, in CLNE allows each player to react optimally to the other players' behavior. This makes it possible for strategic interaction to take place throughout the game's time horizon. Applying OLNE, CLNE, and FBNE in an SC depends on the extent to which chain members have and share state information. If a chain member cannot get such information, it cannot condition its actions on the state vector, and the firm must apply an open-loop strategy. In the context of an SC, an important difference between the decision rules is that feedback and closed-loop strategies both involve the observability of the state variable at any time period while open-loop strategies do not impose such a requirement (e.g., El Ouardighi & Erickson, 2015). In general, there exist differences between OLNE, CLNE, and FBNE strategies for the class of differential games to which our model belongs, that is, linear-quadratic games (Engwerda, 2005). Therefore, in theory, our problem may, in the case of finite time horizon game with symmetric players, such as ours, admit six scenarios: three with homogeneous (OLNE *vs.* OLNE, CLNE *vs.* CLNE, and FBNE *vs.* FBNE strategies) and three with heterogeneous decision patterns (OLNE *vs.* CLNE, OLNE *vs.* FBNE, and CLNE *vs.* FBNE strategies). It is obvious that the static version of our SC game cannot embrace such diversity of decision patterns. Our analysis hereafter will show the discrepancies between the outcomes obtained from the static version, on the one hand, and from two selected scenarios from the dynamic setting, that is, one with homogenous (OLNE *vs.* OLNE strategies) and one with heterogeneous decision patterns (OLNE *vs.* CLNE strategies), on the other hand.

The cooperative objective function in a dynamic setting is formulated as:

$$\begin{aligned}\underset{\pi_s,\pi_m(t),u_s(t),u_m(t)}{\text{Max}} \quad \Pi_c^d = \int_0^T \{(\pi_m(t)+\pi_s)\left[\alpha-\beta\left(\pi_s+\omega_m-\theta X(t)+\pi_m(t)\right)\right] \\ -u(t)^2/2-v(t)^2/2\}\,\mathrm{d}t\end{aligned} \tag{8.74}$$

under the constraint (8.65), where the subscript $c$ stands for cooperative.

For the static setting, the centralized problem is:

$$\begin{aligned}\underset{\pi_s,\pi_m,u_s,u_m}{\text{Max}} \quad \Pi_c^s = \left\{(\pi_m+\pi_s)\left[\alpha-\beta\left(\pi_s+\omega_m-\theta\,(u+v)\,\tfrac{(T-1)}{2}+\pi_m\right)\right]\right. \\ \left.-\tfrac{u^2}{2}-\tfrac{v^2}{2}\right\}T\end{aligned} \tag{8.75}$$

## 8.4.2 *Analysis*

Again, to ensure the feasibility of the solutions and comparability between the results, we assume a similar game duration for the four cases considered, which corresponds here to the shortest time horizon that ranges between the dynamic cooperative setting ($T < \frac{\pi}{2\sqrt{\beta\theta}}$) and the static non-cooperative setting ($T < 1+\frac{\sqrt{6}}{\theta\sqrt{\beta}}$), that is, $T < \frac{\pi}{2\sqrt{\beta\theta}}$, which is granted for $\theta < 1$.

### 8.4.2.1 Cooperative Supply Chain

We first consider the cooperative static solution.

**Lemma 11** The static cooperative profit margins and R&D efforts are given by:

$$\pi^s_{s,c} = \pi^s_{m,c} = \frac{\alpha - \beta\omega_m}{\beta\left[4 - \beta\theta^2(T-1)^2\right]} \tag{8.76}$$

$$u^s_c = v^s_c = \frac{\theta\,(T-1)\,(\alpha - \beta\omega_m)}{4 - \beta\theta^2(T-1)^2} \tag{8.77}$$

with the corresponding cooperative sales:

$$S^s_c = \frac{2\,(\alpha - \beta\omega_m)}{4 - \beta\theta^2(T-1)^2} \tag{8.78}$$

and the maximized cooperative profit:

$$\Pi^s_c = \frac{(\alpha - \beta\omega_m)^2 T}{\beta\left[4 - \beta\theta^2(T-1)^2\right]^2} \tag{8.79}$$

**Proof. A.3.1**
We now consider the cooperative dynamic solution.

**Lemma 12** The dynamic cooperative profit margins and R&D efforts are given by:

$$\pi^d_{s,c} = \pi^d_{m,c} = \left\{1 + \theta\left[\frac{\cos\left[\sqrt{\beta\theta}\,(T-t)\right]}{\cos\left(T\sqrt{\beta\theta}\right)} - 1\right]\right\}\frac{\alpha - \beta\omega_m}{2\beta} \tag{8.80}$$

$$u^d_c = v^d_c = \frac{\sqrt{\theta}\sin\left[\sqrt{\beta\theta}\,(T-t)\right]}{2\sqrt{\beta}\cos\left(T\sqrt{\beta\theta}\right)}\,(\alpha - \beta\omega_m) \tag{8.81}$$

with the corresponding cooperative sales:

$$S^d_c = \left\{1 + \theta\left[\frac{\cos\left[\sqrt{\beta\theta}\,(T-t)\right]}{\cos\left(T\sqrt{\beta\theta}\right)} - 1\right]\right\}\frac{\alpha - \beta\omega_m}{2} \tag{8.82}$$

and the maximized cooperative profit:

$$\Pi^d_c = \left\{\left[\frac{2\cos^2\left(T\sqrt{\beta\theta}\right)(1-\theta)^2(2+\theta) - \theta\left(1-\theta-\theta^2\right)}{\cos^2\left(T\sqrt{\beta\theta}\right)}\right]T + \frac{\sqrt{\theta}\left(9-2\theta-3\theta^2\right)\tan\left(T\sqrt{\beta\theta}\right)}{\sqrt{\beta}}\right\}\frac{(\alpha-\beta\omega_m)^2}{8\beta} \tag{8.83}$$

**Proof. A.3.2**
From (8.80), we observe that the cooperative profit margins increase concavely over time and reach a maximum value at $t = T$. Comparing the cooperative profit margins in (8.76) and (8.80), it is easy to see that $\pi^d_{m,nc}(0) = \pi^d_{s,nc}(0) > \pi^s_{s,c} = \pi^s_{m,c}$ under

$T < \frac{\pi}{2\sqrt{\beta\theta}}$, which, because of non-decreasing dynamic cooperative profit margins over time, implies that $\pi^d_{s,c}(t) = \pi^d_{m,c}(t) > \pi^s_{s,c} = \pi^s_{m,c}$ for any $t \leq T$.

Hence the following proposition:

**Proposition 7** Static decision rules result in lower profit margins than farsighted decision rules in a cooperative SC.

From (8.154), the consumer price is initially equal to:

$$p^d_c(0) = \frac{\alpha + \beta\omega_m}{2\beta} \tag{8.84}$$

and then decreases convexly, while the consumer demand increases concavely. From (8.81), we observe that the SC members' R&D efforts are initially:

$$u^d_c(0) = v^d_c(0) = \frac{\sqrt{\theta}\tan\left(T\sqrt{\beta\theta}\right)}{2\sqrt{\beta}}(\alpha - \beta\omega_m) \tag{8.85}$$

and decreasing over time until they become zero at the end of the game horizon. Relatedly, from (8.153), the stock of induced learning increases concavely over time.

#### 8.4.2.2 Non-cooperative Supply Chain

**Lemma 13** The static Stackelberg equilibrium strategies for the SC members' profit margins and R&D efforts are given by:

$$\pi^s_{s,nc} = \frac{\left[8 - \beta\theta^2(T-1)^2\right](\alpha - \beta\omega_m)}{\beta\left[16 - 3\beta\theta^2(T-1)^2\right]} \tag{8.86}$$

$$\pi^s_{m,nc} = \frac{4(\alpha - \beta\omega_m)}{\beta\left[16 - 3\beta\theta^2(T-1)^2\right]} \tag{8.87}$$

$$u^s_{nc} = v^s_{nc} = \frac{2\theta(T-1)(\alpha - \beta\omega_m)}{16 - 3\beta\theta^2(T-1)^2} \tag{8.88}$$

with the corresponding non-cooperative sales:

$$S^s_{nc} = \frac{4(\alpha - \beta\omega_m)}{16 - 3\beta\theta^2(T-1)^2} \tag{8.89}$$

and the maximized non-cooperative profits:

$$\Pi^s_{s,nc} = \frac{2(\alpha - \beta\omega_m)^2 T}{\beta\left[16 - 3\beta\theta^2(T-1)^2\right]} \tag{8.90}$$

$$\Pi^s_{m,nc} = \frac{2\left[8 - \beta\theta^2(T-1)^2\right](\alpha - \beta\omega_m)^2 T}{\beta\left[16 - 3\beta\theta^2(T-1)^2\right]^2} \tag{8.91}$$

where the subscript *nc* stands for non-cooperative.

**Proof. A.3.3**
From (8.88), it appears that despite a hierarchical mode of play and different profit margins in (8.86)–(8.87), the SC members make equivalent R&D efforts. Finally, from (8.90) and (8.91), the supplier gets a greater overall profit than the manufacturer, which is intuitive.

Considering the dynamic version of the decentralized game model, we assume the possibility of heterogeneous strategies where the manufacturer is committed to efficiency and plays an OLNE strategy, while the supplier may adopt either a contingent (i.e., non-committed) or a committed behavior, that is, either an OLNE or CLNE strategy.

**Lemma 14** The dynamic non-cooperative profit margins and R&D efforts are given by:

$$\pi^d_{s,nc} = \frac{6\tan\left(\theta T\sqrt{\beta/2}\right)(\alpha - \beta\omega_m)}{\beta\left\{(2-h)\theta T\sqrt{\beta/2}\left[(2-h)\beta\theta^2T^2 + 12\right] - 12(1-h)\tan\left(\theta T\sqrt{\beta/2}\right)\right\}} \tag{8.92}$$

$$\begin{aligned}\pi^d_{m,nc} =& \frac{\cos\left[\theta(T-t)\sqrt{\beta/2}\right]}{2\beta\cos\left(\theta T\sqrt{\beta/2}\right)}(\alpha - \beta\omega_m)\\ &- \frac{1}{2}\left\{1 - (1-h)\left[\frac{\cos\left[\theta(T-t)\sqrt{\beta/2}\right]}{\cos\left(\theta T\sqrt{\beta/2}\right)} - 1\right]\right\}\pi^d_{s,nc}\end{aligned} \tag{8.93}$$

$$u^d_{nc}(t) = \frac{(2-h)\beta\theta(T-t)}{2}\pi^d_{s,nc} \tag{8.94}$$

$$\begin{aligned}v^d_{nc}(t) =& \frac{\sin\left[\theta(T-t)\sqrt{\beta/2}\right]\left\{\alpha + \beta\left[(1-h)\pi^d_{s,nc} - \omega_m\right]\right\}}{\sqrt{2\beta}\cos\left(\theta T\sqrt{\beta/2}\right)}\\ &- \frac{(2-h)\beta\theta(T-t)}{2}\pi^d_{s,nc}\end{aligned} \tag{8.95}$$

with the corresponding non-cooperative sales:

$$S^d_{nc} = \frac{\cos\left[\theta\,(T-t)\sqrt{\beta/2}\right]}{\cos\left(\theta T\sqrt{\beta/2}\right)}(\alpha - \beta\omega_m) + \left\{\frac{(1-h)\cos\left[\theta\,(T-t)\sqrt{\beta/2}\right]}{\cos\left(\theta T\sqrt{\beta/2}\right)} - \frac{(2-h)}{2}\right\}\beta\pi^d_{s,nc} \tag{8.96}$$

and the non-cooperative profits:

$$\Pi^d_{s,nc} = \frac{\begin{array}{c}\left\{3\sqrt{2}\tan\left(\theta T\sqrt{\beta/2}\right)\left[\alpha - \beta\left[\omega_m - (1-h)\,\pi^d_{s,nc}\right]\right]\right.\\ \left.-\pi^d_{s,nc}\beta\sqrt{\beta}\theta T\left[3\,(2-h) + \beta\theta^2T^2\right]\right\}\pi^d_{s,nc}\end{array}}{6\sqrt{\beta}\theta} \tag{8.97}$$

$$\Pi^d_{m,nc} = \frac{\begin{array}{c}3\sqrt{2}\tan\left(\theta T\sqrt{\beta/2}\right)\pi^d_{m,nc}\left[\alpha - \beta\left[\omega_m - (1-h)\,\pi^d_{s,nc}\right]\right]\\ -\,\pi^d_{s,nc}\beta\sqrt{\beta}\theta T\left[3\,(2-h)\,\pi^d_{m,nc} + \beta\theta^2T^2\pi^d_{s,nc}\right]\end{array}}{6\sqrt{\beta}\theta} \tag{8.98}$$

where $h = 0$ if the supplier is committed to efficiency, and $h = 1$ in the converse case.

**Proof. A.3.4**
From (8.90), it can be shown that, depending on whether or not the supplier is committed to efficiency, its profit margin can differ significantly (e.g., El Ouardighi & Shnaiderman, 2019). Accordingly, we see from (8.94) that the supplier's effort also differs depending on whether or not the supplier is committed to efficiency. From (8.93), the manufacturer's profit margin is increasing over time and reaches a maximum value at $t = T$. Whereas $\pi^d_{m,h=1}(0) \neq \pi^d_{m,h=0}(0)$, the evolution of the manufacturer's profit margin over time depends on whether or not the supplier is committed to efficiency. Finally, from (8.96), the sales are greater if the supplier is committed to efficiency than in the converse case, so that $S^d_{nc,h=0} > S^d_{nc,h=1}$ $\forall t > 0$. This result leads to the following proposition:

**Proposition 8** The supplier's commitment to efficiency contributes to mitigating the double marginalization effect more effectively than in the converse case.

It is obvious that these differences in behaviors and their subtle implications are totally omitted in the static version of the SC game.

## 8.5 Concluding Remarks

A static, shortsighted policy involves decisions that remain constant and independent of time and the evolving system state throughout a given period. It starkly contrasts dynamic policies that adapt to changing conditions and evolving strategies. From the viewpoint of dynamic game theory scenarios, the inefficacy of static models in dynamic game theory scenarios emanates from their inability to incorporate the temporal dimension of decision-making and grasp the far-reaching consequences these decisions can wield on an ever-shifting landscape. The objective here was to thoroughly explore the challenges and limitations associated with static, shortsighted policies within the dynamic game theory framework.

To achieve optimal performance, supply chains require a strategic balance between short-term and long-term gains. Therefore, integrating long-term perspectives into SC game models is crucial for ensuring both efficiency and effectiveness. Dynamic models, which adapt to evolving conditions, offer more realistic strategies. The analysis demonstrates that dynamic SC game models generally lead to greater cooperation incentives and more effective management of inefficiencies compared to static models.

We expect that this chapter will encourage researchers in supply chain management to opt for dynamic rather than static models, and motivate leading scientific journals in the area of supply chain management to consider publishing articles based on dynamic SC games.

## Appendices

### *Appendix 1*

**A.1.1.** Via a direct optimization of the profit function (8.11) with respect to the decision variable $p$, one gets:

$$\frac{\partial \Pi_c^s}{\partial p} = 0 \Longrightarrow p_c^s = \frac{[1 - \beta\theta\,(T-2)]\,\alpha + \beta\omega_m}{\beta\,[2 - \beta\theta\,(T-2)]} \tag{8.99}$$

where the superscript $s$ stands for static and the subscript $c$ for cooperative. From (8.99), $p_c^s$ is strictly positive if $T < 1 + \frac{2}{\beta\theta}$. Plugging the expression of $p_c^s$ in (8.11) gives (8.14). Then, it is straightforward to obtain (8.13). At the end of the planning horizon, the accumulated experience is given by:

$$X_c^s(T) = \frac{(\alpha - \beta\omega_m)\,(T-1)}{2\,[2 - \beta\theta\,(T-2)]} \tag{8.100}$$

□

**A.1.2.** Using (8.10), (8.1), and (8.2), the Hamiltonian writes (Sethi, 2021)

$$\mathcal{H} = [p - (\omega_m - \theta X) + \lambda] (\alpha - \beta p) \tag{8.101}$$

where $\lambda \equiv \lambda(t)$ is a costate variable.

The necessary condition for optimality is:

$$\mathcal{H}_p = 0 \Longrightarrow p_c^d = \frac{1}{2}\left(\frac{\alpha}{\beta} + \omega_m - \theta X - \lambda\right) \tag{8.102}$$

where the superscript $d$ stands for dynamic and the subscript $c$ for cooperative. From (8.102), we conclude that the model has the linear-quadratic property (Dockner et al., 1985). It can be easily seen that the Hamiltonian is concave in the control variable, $\mathcal{H}_{pp} < 0$. The fact that the state variable has a positive influence on the objective criterion suggests that its corresponding costate variable is positive (Léonard, 1981), i.e., $\lambda(t) \geq 0$. Accordingly, $\lambda(t)$ is interpreted as a marginal incentive to accumulate experience. If $\lambda(t) \geq 0$, the control variable should take on non-negative values and the sales are:

$$S_c^d = \frac{1}{2} [\alpha - \beta (\omega_m - \theta X - \lambda)] \tag{8.103}$$

Using (8.16), we derive the following two-point boundary value problem (Grass et al., 2008):

$$\dot{\lambda} = -\frac{\theta}{2} [\alpha - \beta (\omega_m - \theta X - \lambda)], \quad \lambda(T) = 0 \tag{8.104}$$

$$\dot{X}(t) = \frac{1}{2} [\alpha - \beta (\omega_m - \theta X - \lambda)], \quad X(0) = 0 \tag{8.105}$$

which is solved as:

$$\lambda(t) = \frac{(\alpha - \beta\omega_m)\theta (T - t)}{2 - \beta\theta T} \tag{8.106}$$

$$X_c^d(t) = \frac{(\alpha - \beta\omega_m) t}{2 - \beta\theta T} \tag{8.107}$$

According to (8.105), the cooperative stock of experience starts from $X_c^d(0) = 0$ and increases linearly over time to end up at $X_c^d(T) = \frac{(\alpha-\beta\omega_m)T}{2-\beta\theta T}$. However, the feasibility of the solution requires a sufficiently short planning horizon, that is, $T < 2/\beta\theta$. Using (8.106)–(8.107) for the expression of $p_c^d$ in (8.102) gives (8.15), which is then used in (2) to provide (8.16). Finally, plugging the expression of $X_c^d$

from (107) and the expression of $p_c^d$ from (8.15) into (8.10) and resolving gives (17). □

**A.1.3.** From (8.8), it is straightforward to obtain the manufacturer's static non-cooperative sales price, that is:

$$\frac{\Pi_m^s}{\partial p_m} = 0 \Longrightarrow p_m^s = \frac{[1 - \beta\theta\,(T-2)]\,\alpha + \beta\,(p_s + \omega_m)}{\beta\,[2 - \beta\theta\,(T-2)]} \tag{8.108}$$

which is strictly positive if $T < 1 + \frac{1}{\beta\theta}$. Substituting (8.108) into (8.9) and resolving for the supplier's transfer price gives (8.21), which finally results in the static non-cooperative sales price (8.20). Using the sales price in (8.2) gives the static non-cooperative sales in (8.22). Finally, plugging the expressions of $p_m^s$ and $p_s^s$ from (8.20) and (8.21), respectively, into (8.8) and (8.9) and resolving gives (8.23) and (8.24). At the end of the game, the manufacturer's accumulated experience is given by:

$$X_m^s(T) = \frac{(\alpha - \beta\omega_m)\,(T-1)}{4\,[2 - \beta\theta\,(T-2)]}. \tag{8.109}$$

□

**A.1.4.** From (8.14) and (8.23)–(8.24), we use the Nash bargaining scheme, that is:

$$\Pi_m^{sc} = \Pi_m^s + \frac{\Pi_c^s - \Pi_m^s - \Pi_s^s}{2}$$

$$\Pi_s^{sc} = \Pi_s^s + \frac{\Pi_c^s - \Pi_m^s - \Pi_s^s}{2}$$

to obtain (8.25)–(8.26). □

**A.1.5.** Using (8.6), (8.1), and (8.2), the manufacturer's Hamilton-Jacobi-Bellman (HJB) equation writes:

$$-\dot{V} = \left[p_m - (p_s + \omega_m - \theta X) + \frac{\partial V}{\partial X}\right](\alpha - \beta p_m) \tag{8.110}$$

where $V(X)$ is the manufacturer's value function.

The manufacturer's equilibrium condition is:

$$p_m^d = \frac{1}{2}\left(\frac{\alpha}{\beta} + p_s + \omega_m - \theta X - \frac{\partial V}{\partial X}\right) \tag{8.111}$$

while the sales are given by:

$$S_m^d = \frac{1}{2}\left[\alpha - \beta\left(\omega_m + p_s - \theta X - \frac{\partial V}{\partial X}\right)\right] \tag{8.112}$$

Using (8.111) and (8.112), (8.110) rewrites:

$$-\dot{V} = \frac{1}{4\beta}\left[\alpha - \beta\left(p_s + \omega_m - \theta X - \frac{\partial V}{\partial X}\right)\right]^2. \tag{8.113}$$

We make the following conjecture regarding the manufacturer's value function, that is:

$$V(X) = \frac{A(t)}{2}X^2 + B(t)X + C(t) \tag{8.114}$$

where $A(t)$, $B(t)$, and $C(t)$ are time-dependent coefficients. From (8.114), we get $\frac{\partial V}{\partial X} = AX + B$. Therefore, the manufacturer's HJB equation in (8.113) becomes:

$$\frac{1}{4\beta}\{\alpha - \beta\left[p_s + \omega_m - B - (\theta + A)\,X\right]\}^2 + \frac{\dot{A}}{2}X^2 + \dot{B}X + \dot{C} = 0 \tag{8.115}$$

which, after elementary manipulations, gives the system of ordinary differential equations:

$$\dot{A} = -\frac{\beta}{2}(\theta + A)^2, \qquad A(T) = 0 \tag{8.116}$$

$$\dot{B} = -\frac{1}{2}(\theta + A)\left[\alpha - \beta\left(\omega_m + p_s - B\right)\right], \qquad B(T) = 0 \tag{8.117}$$

$$\dot{C} = -\frac{1}{4\beta}\left[\alpha - \beta\left(\omega_m + p_s - B\right)\right]^2, \qquad C(T) = 0 \tag{8.118}$$

The solution of (8.116)–(8.118) resolves the HJB Eq. (8.115), that is:

$$(A(t), B(t), C(t)) = \left( \frac{\beta\theta^2 (T-t)}{2-\beta\theta (T-t)}, \frac{\theta [\alpha-\beta (\omega_m+p_s)] (T-t)}{2-\beta\theta (T-t)}, -\frac{[\alpha-\beta (\omega_m+p_s)]^2 (T-t)}{2\beta [2-\beta\theta (T-t)]} \right) \tag{8.119}$$

The manufacturer's sales price then rewrites:

$$p_m^d = \frac{[1-\beta\theta (T-t)]\alpha + \beta (\omega_m + p_s - \theta X)}{\beta [2-\beta\theta (T-t)]} \tag{8.120}$$

while the sales are now given by:

$$S_m^d = \frac{\alpha - \beta (\omega_m + p_s - \theta X)}{2-\beta\theta (T-t)} \tag{8.121}$$

Using (8.121), the stock of experience is thus resolved as:

$$X_m^d = \frac{[\alpha - \beta (\omega_m + p_s)] t}{2-\beta\theta T} \tag{8.122}$$

where the manufacturer's experience accumulates linearly.

This leads to a constant manufacturer's sales price, that is:

$$p_m^d = \frac{\alpha (1-\beta\theta T) + \beta (\omega_m + p_s)}{\beta (2-\beta\theta T)} \tag{8.123}$$

The supplier's profit then writes:

$$\Pi_s^d = \frac{p_s [\alpha - \beta (\omega_m + p_s)] T}{2-\beta\theta T}$$

which is strictly concave in $p_s$. Therefore, the supplier's equilibrium wholesale price is derived to obtain (8.28). Plugging the expression of $p_s^d$ from (8.28) into (8.123) gives (8.27) while the sales reduce to (8.29), which are positive if $T < 2/\beta\theta$. Using (8.29) in (8.122), the non-cooperative stock of experience becomes:

$$X_m^d = \frac{(\alpha - \beta\omega_m) t}{2 (2-\beta\theta T)} \tag{8.124}$$

Finally, using (8.27), (8.28), (8.29), and (8.124), the manufacturer's and the supplier's profits are obtained in (8.30)–(8.31). □

**A.1.6.** From (8.17) and (8.30)–(8.31), we use the Nash bargaining scheme, that is:

$$\Pi_m^{dc} = \Pi_m^d + \frac{\Pi_c^d - \Pi_m^d - \Pi_s^d}{2}$$

$$\Pi_s^{dc} = \Pi_s^d + \frac{\Pi_c^d - \Pi_m^d - \Pi_s^d}{2}$$

to obtain (8.32)–(8.33). □

## *Appendix 2*

**A.2.1.** We derive the first-order conditions for the cooperative solution in the static setting:

$$\frac{\partial \pi_c^s}{\partial q} = a - 2q - \omega_m + \theta q\,(t-1) = 0 \tag{8.125}$$

which results in the static production rate in (8.9). Plugging the expressions of $q_c^s$ into (8.2) and (8.8), respectively, gives (8.10) and (8.11). Note that the profit function (8.8) is strictly concave in the control variable $q$. □

**A.2.2.** Considering the dynamic setting and skipping the time index for convenience, the Hamiltonian writes

$$\mathcal{H} = [a - q(t) - (\omega_m - \theta X(t))]\,q(t) + \lambda\,(q - \delta X) \tag{8.126}$$

where $\lambda \equiv \lambda(t)$ is a costate variable, given by:

$$\dot{\lambda} = (r + \delta)\,\lambda - \theta q \tag{8.127}$$

Necessary conditions for optimality are:

$$\mathcal{H}_q = a - 2q - (\omega_m - \theta X) + \lambda = 0 \tag{8.128}$$

It is obvious that the Hamiltonian is jointly concave in the control $q$.

From (8.128), the cooperative output is:

$$q_c^d = \frac{a - (\omega_m - \theta X) + \lambda}{2} \tag{8.129}$$

Using (8.129) in (8.127) and (8.1), we get the canonical system:

$$\dot{\lambda} = \left(r + \delta - \frac{\theta}{2}\right)\lambda - \frac{\theta}{2}\left(a - \omega_m + \theta X\right) \tag{8.130}$$

$$\dot{X} = \frac{1}{2}\left(a - \omega_m + \lambda\right) - \left(\delta - \frac{\theta}{2}\right)X \tag{8.131}$$

which is solved for the steady state as:

$$\lambda^{ss} = \frac{\delta\theta\left(a - \omega_m\right)}{2\delta\left(r + \delta\right) - \theta\left(r + 2\delta\right)} \tag{8.132}$$

and (8.13), where the superscript $ss$ stands for steady state. The steady state is feasible if the sufficient condition $\delta \geq \theta$ holds. Therefore, the steady state production rate is given by (8.12). Note that the transversality condition:

$$\lim_{t\to\infty} e^{-rt}\lambda(t)X(t) = 0 \tag{8.133}$$

holds. The stability of the steady state is characterized by the trace and the determinant of the Jacobian matrix of the canonical system:

$$J = \begin{bmatrix} r + \delta - \frac{\theta}{2} & -\frac{\theta^2}{2} \\ \frac{1}{2} & -\delta + \frac{\theta}{2} \end{bmatrix}$$

that is,

$$Tr\ J = r > 0$$

$$|J| = -\frac{1}{2}\left[2\delta\left(r + \delta\right) - \theta\left(r + 2\delta\right)\right]$$

which is negative if (and only if) $\theta < \frac{2\delta(r+\delta)}{r+2\delta}$, with the eigenvalues:

$$\frac{r}{2} \pm \sqrt{\frac{r^2}{4} + \frac{1}{2}\left[2\delta\left(r + \delta\right) - \theta\left(r + 2\delta\right)\right]}$$

one being positive and one negative under the condition $\theta < \frac{2\delta(r+\delta)}{r+2\delta}$. Therefore, the steady state has the saddle-point property. In the converse case where $\theta \geq \frac{2\delta(r+\delta)}{r+2\delta}$, there is no feasible steady state.

Finally, using (8.12)–(8.13) in (8.7), the cooperative steady state profit is computed in (8.14). □

**A.2.3.** Based on the manufacturer's and supplier's profit functions in (8.5) and (8.6), respectively, it is straightforward to obtain the manufacturer's static non-cooperative equilibrium output, that is:

$$\frac{\pi_m^s}{\partial q} = (a - p_s - \omega_m) - 2q\left[1 - \theta\left(T - 1\right)\right] = 0 \tag{8.134}$$

and the supplier's static non-cooperative transfer price, that is:

$$\frac{\pi_s^s}{\partial p_s} = \frac{a - \omega_m}{2\left[1 - \theta\left(T - 1\right)\right]} - \frac{2p_s}{2\left[1 - \theta\left(T - 1\right)\right]} = 0 \tag{8.135}$$

Resolving for $p_s^s$ and $q_{nc}^s$, we get (8.15), (8.16). The SC members' profits are finally computed as (8.17) and (8.18).

**A.2.4.** We now turn to the non-cooperative dynamic setting and confine our interest to equilibrium outcomes for which the objective integrals in (8.3) and (8.4) converge for all admissible states, controls, and parameter values

Skipping the time index for convenience, the manufacturer's HJB equation writes:

$$rV^m = \left[a - q - p_s - (\omega_m - \theta X)\right]q + \frac{\partial V^m}{\partial X}\left(q - \delta X\right) \tag{8.136}$$

where $V^m(X)$ is the manufacturer's value functions.

The manufacturer's equilibrium condition is:

$$q = \frac{1}{2}\left[a - p_s - (\omega_m - \theta X) + \frac{\partial V^m}{\partial X}\right] \tag{8.137}$$

Plugging (8.137) into (8.136) gives:

$$rV^m = \frac{1}{4}\left[a - p_s - (\omega_m - \theta X) + \frac{\partial V^m}{\partial X}\right]^2 - \frac{\partial V^m}{\partial X}\delta X \tag{8.138}$$

Let us make the following conjecture:

$$V^m(X) = \frac{A}{2}X^2 + BX + C \tag{8.139}$$

from which $\frac{\partial V^m}{\partial X} = AX + B$. After substitution in the HJB equation, we get:

$$\frac{1}{4}(a - p_s - \omega_m + B)^2 - rC + \left[\frac{1}{2}(a - p_s - \omega_m + B)(\theta + A) - (r + \delta) B\right] X$$
$$+ \left[\frac{1}{4}(\theta + A)^2 - \left(\frac{r}{2} + \delta\right) A\right] X^2 = 0 \tag{8.140}$$

Two solutions resolve (8.140), but only one ensures the condition of convergence of the state Eq. (8.1) under $\theta < \frac{2\delta(r+\delta)}{r+2\delta}$, that is,

$$(A, B, C) = \left(r+2\delta-\theta-\chi, \frac{(r + 2\delta-\chi)(a-p_s-\omega_m)}{r+\chi}, \frac{(r+\delta)^2(a-p_s-\omega_m)^2}{2r\left\{(r+\delta)^2+\delta^2 - \theta(r+2\delta) + r\chi\right\}}\right) \tag{8.141}$$

where $\chi = \sqrt{(r + 2\delta)(r + 2\delta - 2\theta)}$.

Using (8.141) to resolve (8.1), and assuming $X_0 = 0$ for simplicity, we get the globally asymptotically stable solution, that is,

$$X_{nc}^d(t) = \frac{(r + \delta)(a - p_s - \omega_m)}{2\delta(r + \delta) - \theta(r + 2\delta)}\left[1 - e^{-\frac{[2\delta(r+\delta)-\theta(r+2\delta)]t}{r+\chi}}\right] \tag{8.142}$$

which results in the steady state value in (8.21). Note that the learning effect in (8.142) is increasing concavely over time. Using (8.141) and (8.142) to resolve (8.137) and using the resulting expression to resolve (8.4), we derive the optimal supplier's transfer price in (8.19). Finally, the manufacturer's and the supplier's steady state profits are obtained in (8.22)–(8.23). □

## *Appendix 3*

**A.3.1.** From (8.11), letting $\pi \equiv \pi_s + \pi_m$, it is straightforward to obtain:

$$\frac{\partial \Pi_c^s}{\partial \pi} = 0 \Longrightarrow \pi_c^s = \frac{1}{2\beta}\left\{\alpha - \beta\left[\omega_m - \theta(u + v)\frac{(T - 1)}{2}\right]\right\} \tag{8.143}$$

$$\frac{\partial \Pi_c^s}{\partial u} = \frac{\partial \Pi_c^s}{\partial v} = 0 \Longrightarrow u_c^s = v_c^s = \beta\theta \frac{(T-1)}{2}\pi \tag{8.144}$$

where the subscript $c$ stands for cooperative, which results in the static cooperative decision rules in (8.12)–(8.13). Note that (8.12)–(8.13) are strictly positive for any $T < 1+\frac{2}{\theta\sqrt{\beta}}$. It can be easily shown that the profit function is jointly concave in the control vector $\left(\pi_c^s, u_c^s, v_c^s\right)$ because the corresponding Hessian matrix is negative definite.

Then, the corresponding sales and the optimal cooperative profit, which are strictly positive for any $T < 1 + \frac{2}{\theta\sqrt{\beta}}$, are given in (8.14)–(8.15). □

**A.3.2.** Turning to the dynamic setting, and using $\boldsymbol{\pi}(t) \equiv \boldsymbol{\pi}_s + \boldsymbol{\pi}_m(t)$, the Hamiltonian writes:

$$\mathcal{H} = \pi\left[\alpha - \beta\left(\omega_m - \theta X + \pi\right)\right] - \frac{u^2}{2} - \frac{v^2}{2} + \lambda\left(u + v\right) \tag{8.145}$$

where $\lambda \equiv \lambda(t)$ is a costate variable and the time index is skipped for convenience.

The costate equation is given by:

$$\dot{\lambda} = -\beta\theta\pi \tag{8.146}$$

with the transversality condition $\lambda(T) = 0$.

Necessary conditions for optimality are:

$$\mathcal{H}_\pi = 0 \Rightarrow \pi_c^d = \frac{1}{2\beta}\left[\alpha - \beta\left(\omega_m - \theta X\right)\right] \tag{8.147}$$

$$\mathcal{H}_u = \mathcal{H}_v = 0 \Rightarrow u_c^d = v_c^d = \lambda \tag{8.148}$$

It can be easily seen that the Hamiltonian is jointly concave in the control vector $(\pi, u, v)$ since the corresponding Hessian matrix is negative definite.

Using (8.147), the cooperative consumer demand is:

$$S_c^d = \frac{1}{2}\left[\alpha - \beta\left(\omega_m - \theta X\right)\right] \tag{8.149}$$

Plugging the expressions from (8.147) and (8.148), respectively, into (8.146) and (8.1), we get the TPBVP:

$$\dot{\lambda} = -\frac{\theta}{2}\left[\alpha - \beta\left(\omega_m - \theta X\right)\right] \quad \lambda(T) = 0 \tag{8.150}$$

$$\dot{X} = 2\lambda, \qquad X_0 \geq 0 \tag{8.151}$$

which, for $X_0 = 0$, is solved as:

$$\lambda(t) = \frac{\sqrt{\theta}\sin\left[\sqrt{\beta\theta}\,(T-t)\right]}{2\sqrt{\beta}\cos\left(T\sqrt{\beta\theta}\right)}(\alpha - \beta\omega_m) \tag{8.152}$$

$$X_c^d(t) = \frac{1}{\beta}\left\{\frac{\cos\left[\sqrt{\beta\theta}\,(T-t)\right]}{\cos\left(T\sqrt{\beta\theta}\right)} - 1\right\}(\alpha - \beta\omega_m) \tag{8.153}$$

where $T < \frac{\pi}{2\sqrt{\beta\theta}}$ is required for non-negative solutions.

From (8.147) and (8.152), we get (8.17). The cooperative profit margins are then given by (8.16). From (8.16) and (8.153), the consumer price is:

$$p_c^d = \frac{\begin{array}{c}\left\{(1+\theta)\cos\left(T\sqrt{\beta\theta}\right) - \theta\cos\left[\sqrt{\beta\theta}\,(T-t)\right]\right\}\alpha \\ +\left\{(1-\theta)\cos\left(T\sqrt{\beta\theta}\right) + \theta\cos\left[\sqrt{\beta\theta}\,(T-t)\right]\right\}\beta\omega_m\end{array}}{2\beta\cos\left(T\sqrt{\beta\theta}\right)} \tag{8.154}$$

from which the demand is given in (8.18).

From (8.153), at the end of the planning horizon, the manufacturer's operating cost is such that:

$$\omega_m - \theta X_c^d(T) = -\frac{1}{\beta}\left\{\frac{1}{\cos\left(T\sqrt{\beta\theta}\right)} - 1\right\}\left[\theta\alpha - (1+\theta)\,\beta\omega_m\right] > 0 \mid \frac{\alpha}{\beta\omega_m} < \frac{1+\theta}{\theta}$$

Finally, for $T < \frac{\pi}{2\sqrt{\beta\theta}}$, the cooperative cumulative profit is given in (8.19), which is clearly positive. □

**A.3.3.** In the static setting, the Nash equilibrium conditions are:

$$\frac{\partial \Pi_{s,nc}^s}{\partial \pi_{s,nc}^s} = 0 \Longrightarrow \pi_{s,nc}^s = \frac{1}{2\beta}\left\{\alpha - \beta\left[\omega_m + \pi_{m,nc}^s - \theta\,(u+v)\,\frac{(T-1)}{2}\right]\right\}T \tag{8.155}$$

$$\frac{\partial \Pi_{s,nc}^s}{\partial u_{nc}^s} = 0 \Longrightarrow u_{nc}^s = \beta\theta\frac{(T-1)}{2}\pi_{m,nc}^s \tag{8.156}$$

$$\frac{\partial \Pi^s_{m,nc}}{\partial \pi^s_{m,nc}} = 0 \Longrightarrow \pi^s_{m,nc} = \frac{1}{2\beta}\left\{\alpha - \beta\left[\omega_m + \pi^s_{s,nc} - \theta\,(u+v)\,\frac{(T-1)}{2}\right]\right\} \tag{8.157}$$

$$\frac{\partial \Pi^s_{m,nc}}{\partial v^s_{nc}} = 0 \Longrightarrow v^s_{nc} = \beta\theta\frac{(T-1)}{2}\pi^s_{s,nc} \tag{8.158}$$

where the subscript *nc* stands for non-cooperative, which results in the static cooperative decision rules in (8.12)–(8.13). Note that (8.12)–(8.14) are strictly positive for any $T < 1 + \frac{\sqrt{6}}{\theta\sqrt{\beta}}$. Here also, it can be easily shown that each player's profit function is jointly concave in the corresponding control vector, i.e., $\left(\pi^s_s, u^s_{nc}\right)$ for the supplier and $\left(\pi^s_m, v^s_{nc}\right)$ for the manufacturer. The corresponding sales and profits, which are strictly positive for any $T < 1 + \frac{\sqrt{6}}{\theta\sqrt{\beta}}$, are given in (8.15)–(8.16). □

**A.3.4.** In the dynamic context, the SC members' Hamiltonians write:

$$\mathcal{H}^s = \pi_s\left[\alpha - \beta\left(\pi_s + \omega_m - \theta X + \pi_m\right)\right] - \frac{u^2}{2} + \lambda^s\,(u+v) \tag{8.159}$$

$$\mathcal{H}^m = \pi_m\left[\alpha - \beta\left(\pi_s + \omega_m - \theta X + \pi_m\right)\right] - \frac{v^2}{2} + \lambda^m\,(u+v) \tag{8.160}$$

where $\lambda^s(t)$ and $\lambda^m(t)$, respectively, denote the supplier's and the manufacturer's costate variable, which interprets as the marginal incentive for efficiency. In this setup, we solve a one-stage game.

The SC members' non-cooperative equilibrium conditions are:

$$\mathcal{H}^m_{\pi_m} = 0 \Rightarrow \pi^d_m = \frac{1}{2\beta}\left[\alpha - \beta\left(\pi_s + \omega_m - \theta X\right)\right] \tag{8.161}$$

$$\mathcal{H}^s_u = 0 \Rightarrow u^d = \lambda^s \tag{8.162}$$

$$\mathcal{H}^m_v = 0 \Rightarrow v^d = \lambda^m \tag{8.163}$$

It can be easily seen that each firm's Hamiltonian is jointly concave in its corresponding control vector.

The SC members' respective costate equations are given by:

$$\dot{\lambda}^s = -\mathcal{H}_X^s - h\left(\mathcal{H}_{\pi_m}^s \frac{\partial \pi_m^d}{\partial X} + \mathcal{H}_v^s \frac{\partial v^d}{\partial X}\right) = -\left(\frac{2-h}{2}\right)\beta\theta\pi_s \tag{8.164}$$

$$\dot{\lambda}^m = -\mathcal{H}_X^m = -\frac{\theta\left[\alpha - \beta\left(\pi_s + \omega_m - \theta X\right)\right]}{2} \tag{8.165}$$

with the transversality conditions $\lambda^s(T) = \lambda^m(T) = 0$. In (8.164)–(8.165), $h = 0$ if the supplier is committed to efficiency and $h = 1$ if the supplier behaves in a contingent way.

Using (8.162)–(8.163), the state equation rewrites:

$$\dot{X} = \lambda^s + \lambda^m, \tag{8.166}$$

Solving the TPBVP composed of (8.164)–(8.166) for $X_0 = 0$, we get:

$$\lambda^s(t) = \left(\frac{2-h}{2}\right)\beta\theta\pi_s(T-t) \tag{8.167}$$

$$\lambda^m(t) = \frac{\sin\left(\theta(T-t)\sqrt{\beta/2}\right)}{\sqrt{2\beta}\cos\left(\theta T\sqrt{\beta/2}\right)}\{\alpha - \beta[\omega_m + (1-h)\pi_s]\} - \left(\frac{2-h}{2}\right)\beta\theta\pi_s(T-t) \tag{8.168}$$

$$X_{nc}^d(t) = \frac{1}{\beta\theta}\left\{\frac{\cos\left[\theta\sqrt{\beta/2}(T-t)\right]}{\cos\left(\theta T\sqrt{\beta/2}\right)} - 1\right\}\{\alpha - \beta[\omega_m + (1-h)\pi_s]\} \tag{8.169}$$

where $T < \frac{\pi}{2\theta\sqrt{\beta/2}}$ is required for non-negative solutions. Using (8.161), (8.169), and (8.168) in (8.6), we derive an optimal value of $\pi_s$ in (8.26) which gives the control time paths in (8.28)–(8.29). Using (8.26) and (8.161), we get (8.27). Using (8.26), (8.27), and (8.169), the non-cooperative consumer demand is obtained in (8.30). Finally, the SC members' non-cooperative overall profits are given by (8.31)–(8.32). □

## References


Agi, M. A. N., Faramari-Oghani, S., & Hazır, Ö. (2021). Game theory-based models in green supply chain management: A review of the literature. *International Journal of Production Research, 59*(15), 4736–4755.

Başar, T., & Olsder, G.-J. (1982). *Dynamic noncooperative game theory*. Academic Press.

Başar, T., & Olsder, G.-J. (1999). *Dynamic non-cooperative game theory* (2nd ed.). SIAM Classics.

Cachon, G. (2001). Stock wars: inventory competition in a two echelon supply chain. *Operations Research, 49*(5), 658–674.

Cachon, G. P. (2003). Supply chain coordination with contracts. In S. C. Graves & A. G. de Kok (Eds.), *Handbooks in operations research and management science*. Springer.

Cachon, G. P., & Fischer, M. (2000). Supply chain inventory management and the value of shared information. *Management Science, 46*(8), 1032–1048.

Cachon, G. P., & Netessine, S. (2004). Game theory in supply chain analysis. In D. Simchi-Levi, D. S. Wu, & M. Zuo-Jun (Eds.), *Handbooks of quantitative supply chain analysis*. Springer.

Chen, F. (2003). Information sharing and supply chain coordination. In S. C. Graves & A. G. de Kok (Eds.), *Handbooks in operations research and management science*. Springer.

Corbett, C. J., & Karmarkar, U. S. (2001). Competition and structure in serial supply chains with deterministic demand. *Management Science, 47*(7), 966–978.

Desai, V. S. (1992). Marketing-production decisions under independent and integrated channel structure. *Annals of Operations Research, 34*(1), 275–306.

Desai, V. S. (1996). Interactions between members of a marketing-production channel under seasonal demand. *European Journal of Operational Research, 90*(1), 115–141.

Dockner, E., Feichtinger, G., & Jørgensen, S. (1985). Tractable classes of nonzero-sum open-loop Nash differential games: Theory and examples. *Journal of Optimization Theory and Applications, 45*(2), 179–197.

Dockner, E., Jørgensen, S., Long, N. V., & Sorger, G. (2000). *Differential games in economics and management science*. Cambridge University Press.

El Ouardighi, F. (2014). Supply quality management with optimal wholesale price and revenue-sharing contracts: A two-stage game approach. *International Journal of Production Economics, 156*(1), 260–268.

El Ouardighi, F., & Erickson, G. (2015). Production capacity buildup and double marginalization mitigation in a dynamic supply chain. *Journal of the Operational Research Society, 66*(8), 1281–1296.

El Ouardighi, F., & Shnaiderman, M. (2019). Supplier's opportunistic behavior and the quality-efficiency tradeoff with conventional supply chain contracts. *Journal of the Operational Research Society, 70*(11), 1915–1937.

El Ouardighi, F., Jørgensen, S., & Pasin, F. (2008). A dynamic game of operations and marketing management in a supply chain. *International Game Theory Review, 10*(4), 373–397.

El Ouardighi, F., Jørgensen, S., & Pasin, F. (2013). A dynamic game with monopolist manufacturer and price-competing retailers. *OR Spectrum, 35*(4), 1059–1084.

El Ouardighi, F., Schnaiderman, M., & Pasin, F. (2014). R&D with stock-dependent spillovers and price competition in a dynamic duopoly. *Journal of Optimization Theory and Applications, 161*(2), 626–647.

El Ouardighi, F., Sim, J. E., & Kim, B. (2016). Pollution accumulation and abatement policy in a supply chain. *European Journal of Operational Research, 248*(3), 982–996.

Eliashberg, J., & Steinberg, R. (1987). Marketing-production decisions in an industrial channel of distribution. *Management Science, 33*(8), 981–1000.

Engwerda, J. (2005). *LQ dynamic optimization and differential games*. Wiley.

Grass, D., Caulkins, J., Feichtinger, G., Tragler, G., & Behrens, D. (2008). *Optimal control of nonlinear processes*. Springer.

Han, J., Li, X., Sethi, S. P., Siu, C. C., & Yam, S. C. P. (2024). Production management with general demands and lost sales. *Operations Research*. https://doi.org/10.1287/opre.2022.0191

He, X., Prasad, A., Sethi, S. P., & Gutierrez, G. J. (2007). A survey of Stackelberg differential game models in supply and marketing channels. *Journal of Systems Sciences and Systems Engineering, 16*, 385–413.

Jørgensen, S. (1986). Optimal production, purchasing and pricing: A differential game approach. *European Journal of Operational Research, 24*(1), 64–76.

Kennedy, A., Sethi, S. P., Siu, C. C., & Chi, S. (2021). Co-op advertising, pricing, and logistics in a dynamic three-echelon supply chain. *Production and Operations Management, 30*(11), 3881–3905.

Kogan, K., & El Ouardighi, F. (2019). Autonomous and induced production learning under price and quality competition. *Applied Mathematical Modelling, 67*, 74–84.

Kogan, K., & Tapiero, C. S. (2007). *Supply chain games: Operations management and risk valuation*. Springer.

Lambertini, L. (2014). Coordinating static and dynamic supply chains with advertising through two-part tariffs. *Automatica, 50*(2), 565–569.

Léonard, D. (1981). The signs of the co-state variables and sufficiency conditions in a class of optimal control problems. *Economics Letters, 8*(4), 321–325.

Long, N. V. (2010). *A survey of dynamic games in economics*. World Scientific.

Nagarajan, M., & Sošić, G. (2008). Game-theoretic analysis of cooperation among supply chain agents: Review and extensions. *European Journal of Operational Research, 187*(3), 719–745.

Sethi, S. P. (2021). *Optimal control theory – Applications to management science and economics* (4th ed.). Springer.

Zaccour, G. (2008). On the coordination of dynamic marketing channels and two-part tariffs. *Automatica, 44*(5), 1233–1239.

Zhao, Y., Wang, S., Cheng, T., Yang, X., & Huang, Z. (2010). Coordination of supply chains by option contracts: A cooperative game theory approach. *European Journal of Operational Research, 207*(2), 668–675.

# Chapter 9
# On the Rebound Effect of Cleaner Technologies and Climate Change: Radical Technology Innovations Needed

**Hassan Benchekroun and Amrita Ray-Chaudhuri**

**Abstract** Technological innovations that reduce emissions per output can backfire and may result in countries increasing their emissions. In the case of climate change, assessing the size of this rebound effect requires a fully fledged dynamic analysis since the externality occurs across space and time. Indeed, greenhouse gas (GHGs) emissions are not only transboundary, their persistence in the atmosphere implies that today's emissions adversely affect current *and* future generations. The welfare analysis needs to account for the sum of all generations' welfare. In a dynamic game, the impact of a technological innovation on emissions is ambiguous and depends on the initial stock of pollution. Therefore, relying on a simplified static version of the game or focusing the analysis on the steady state only can be misleading. In the case of climate change, this rebound effect may be strong enough to result in a decrease of welfare. This perverse effect happens for an empirically relevant range of parameters. Our findings advocate for (i) the necessity of a global agreement on mitigating emissions to accompany the implementation of clean technologies and (ii) policies aimed at fostering research and development in innovative clean technologies to target R&D projects on radical technological innovations rather than targeting a wide range of projects with modest objectives.

**Keywords** Differential games · Radical technological innovation · Climate change · Clean technologies · R&D and climate change

**JEL Classifications** Q20, Q54, Q55, Q58, C73

H. Benchekroun (✉)
Department of Economics, McGill University, Montreal, QC, Canada
e-mail: hassan.benchekroun@mcgill.ca

A. Ray-Chaudhuri
Department of Economics, University of Winnipeg, Winnipeg, MB, Canada

TILEC, Tilburg University, Tilburg, The Netherlands
e-mail: a.raychaudhuri@uwinnipeg.ca

 
F. El Ouardighi, G. Feichtinger (eds.), *The Unaffordable Price of Static Decision-making Models*, International Series in Operations Research & Management Science 365, https://doi.org/10.1007/978-3-031-88638-6_9

## 9.1 Introduction

We examine the impact of implementing clean technologies on levels of emission and welfare in the presence of an accumulative transboundary pollutant. Investment in developing and implementing clean technologies by public and private sectors have steadily increased over the past decades. According to Bloomberg, global clean energy spending has surged by 17% to a record $1.8 trillion in 2023. These include investments to install renewable energy, buy electric vehicles, build hydrogen production systems, and deploy other technologies.[1] In the United States (USA), during 2021–2022, the signing into law of the Investing in America agenda has resulted in the largest investment in reducing carbon emissions in American history (primarily through the Bipartisan Infrastructure Law and the Inflation Reduction Act).[2] Private companies have announced over half a trillion dollars in new investment, including nearly $360 billion in clean energy manufacturing, electric vehicles (EVs) and batteries, and power generation. In the EU, since March 2023, the Commission has approved member state schemes for a total budget of around € 6.9 billion for investment in clean technologies.[3] China has emerged as the world leader with $890bn investment in clean energy sectors in 2023.[4] International organizations, such as the United Nations (UN) and G8, have also proactively encouraged countries to fund the development of clean technologies. Under the UNFCCC, the development and transfer of climate technologies to developing countries is conducted by the Technology Mechanism which was established by Parties in 2010.[5]

We use Benchekroun and RayChaudhuri (2014) to examine the impact of adopting cleaner technologies within a framework that considers transboundary pollution emissions and where pollution emissions accumulate into a stock and therefore inflict lasting damage on the environment,[6] two features characterizing the climate change problem. Considering a world made of $n$ countries or regions, we determine the non-cooperative emissions policies of each region and determine the impact of having all countries simultaneously adopt a cleaner technology (captured

[1] See https://www.bloomberg.com/news/articles/2024-01-30/china-leads-global-clean-energy-spending-which-record-1-8-trillion-in-2023.

[2] These include incentives for manufacturing across the clean energy supply chain, investments in demonstration projects, loans and loan guarantees for a variety of technologies, and production and investment tax credits for clean energy generation.

See https://www.whitehouse.gov/briefing-room/blog/2023/12/19/building-a-thriving-clean-energy-economy-in-2023-and-beyond/.

[3] See https://ec.europa.eu/commission/presscorner/detail/en/ip_23_5245.

[4] See https://www.carbonbrief.org/analysis-clean-energy-was-top-driver-of-chinas-economic-growth-in-2023/.

[5] See https://unfccc.int/topics/adaptation-and-resilience/groups-committees/adaptation-committee/joint-ac-and-leg-mandates/nap-support/technology-development-and-transfer.

[6] See Jørgensen et al. (2010) for a survey of dynamic game models used to analyze environmental problems.

by a decrease in their emission to output ratio). The main findings are illustrated using a numerical example based on updated parameter values that are relevant for climate change, and compared to the findings of Benchekroun and RayChaudhuri (2014).

Since the adoption of a cleaner technology reduces the marginal cost of production (measured in terms of pollution damages), it provides an incentive to each country to increase its production. We find that the increase in emissions associated with the increase in production can outweigh the positive environmental impact of adopting a "cleaner" technology. This is similar to the "rebound effect" found in the literature on energy efficiency whereby energy savings are mitigated when efficiency is improved (see, e.g., Greening et al. 2000; Sorrell & Dimitropoulos 2008). Within a non-cooperative setting, the positive shock of implementing a cleaner technology results in more "aggressive" behavior of countries which ultimately exacerbates the tragedy of the commons. The existence of a rebound effect has also been established in a related contribution Chenavaz et al. (2021). While they do not explicitly consider a pollution problem, they examine the impact of an increase in eco-efficiency. In their model eco-efficiency is a state variable that captures the amount of resources used to produce an output. This impacts the cost of production of the output as well as the demand for the good. Consumers are assumed to have a preference for more eco-efficient products: consumers' maximum willingness to pay is inversely related to the eco-efficiency. The good is supplied by a profit-maximizing monopolist whose optimal control problem consists of maximizing its discounted sum of profits, during an exogenous finite period of time, by controlling the paths of production and investment in eco-efficiency. The authors characterize conditions under which an increase in eco-efficiency can result not only in a decrease in the price charged by the monopolist but also an increase in the overall quantity of resources used; hence, what can be commonly perceived as a positive shock to the industry can backfire and end up in an increase of resource use.

We apply the model presented in Benchekroun and RayChaudhuri (2014) which is based on the seminal transboundary pollution game model in Dockner and Long (1993) and van der Ploeg and de Zeeuw (1992). In contrast with van der Ploeg and de Zeeuw (1992) and Jørgensen and Zaccour (2001), Benchekroun and RayChaudhuri (2014) take the ratio of emissions to output as exogenously given. This captures situations where a cleaner technology is readily available in the more advanced country. van der Ploeg and de Zeeuw (1992) (section 8) and Jørgensen and Zaccour (2001) consider the case where the ratio of emissions to output is endogenous and is a decreasing function of the level of the stock of clean technology. While van der Ploeg and de Zeeuw (1992) assume that the stock of clean technology is public knowledge, Jørgensen and Zaccour (2001) consider the case where the stock of clean technology, also referred to as the stock of abatement capital, is country specific. Each country can invest in the abatement capital in addition to its control of emissions.[7] In line with Benchekroun and RayChaudhuri (2014), we consider exogenously given levels of ratios of emissions to output in order to focus on the existence and implementation of a new technology only and abstract from the game of investment in technologies. While we present the case of exogenous changes in technology, the game we consider can be viewed as a second stage of a two-stage game where in an initial phase countries invest in their technologies. The cleaner technology can be interpreted as an exogenous perturbation of the equilibrium technology choice from the initial stage in investment in technologies. The fact that implementing a technology may have counterintuitive effects is even more striking in our setting, where the new technology is readily available and free.[8] Our conclusions definitely suggest that incentives to invest in abatement technologies need to be reevaluated in the face of non-cooperative emission strategies being implemented by countries.[9]

The main policy recommendation that can be taken from this analysis is that developing cleaner technologies cannot be a substitute for the difficult task of agreeing on and enforcing emission restraints internationally. Moreover, in general in a dynamic game, the impact of a technological innovation on emissions turns out to depend on the level of stock at the time the innovation occurs. The impact is ambiguous and the short-run impact of the technological improvement can be the opposite of its long-run impact. Therefore, to assess accurately the impact of a technological innovation, it is crucial to adopt a dynamic framework since a static model cannot capture such ambiguity.

For completeness, we provide a description of the model used, the Markov perfect Nash equilibrium as well as the analytical analysis of the impact of the adoption of a cleaner technology established in Benchekroun and RayChaudhuri (2014) in respectively Sections 2, 3, and 4. We offer a numerical analysis based

---

[7] van der Ploeg and de Zeeuw (1992) compare the outcome under international policy coordination and the open loop equilibrium when there is no coordination. They show that the level of production and the stock of clean technology are both higher under the non-cooperative equilibrium.

Jørgensen and Zaccour (2001) consider an asymmetric game where there exist two regions facing a pure downstream problem. They design a transfer scheme that induces the cooperative levels of abatement and satisfies overall individual rationality for both regions.

Other papers such as Langinier and RayChaudhuri (2020) focus on firm-level decisions regarding investment in clean innovation.

[8] Another paper to allow for exogenous technology changes within the context of a dynamic model of global warming is Dutta and Radner (2006). Their model differs from ours in the following ways. They model pollution damage as being linear in the stock of pollution, whereas we have a damage function that is strictly convex in the pollution stock. They model a cleaner technology as a reduction in the ratio of emission to input of energy into the production process, whereas we model a cleaner technology as a reduction in the ratio of emission to output. They find that a cleaner technology always increases equilibrium welfare, in contrast to our main result.

[9] We note that there exists a related literature on the "green paradox," where green policies are shown to possibly result in an overshooting of the stock of pollution (see, e.g., Gerlagh 2011; Hoel 2011; Quentin Grafton et al. 2012; van der Ploeg and Withagen 2012; Sinn 2012). In the "green paradox" literature, this result arises due to the impact of such policies on the timing of extraction of a polluting exhaustible resource. The intuition driving the main result in this paper is different since it analyzes a renewable resource.

on empirical evidence of the model parameters in Section 5. Section 6 offers concluding remarks.

## 9.2 The Model

Consider $n$ countries indexed by $i = 1, \ldots, n$. The objective of country $i$ is to maximize its discounted sum of welfare

$$\max_{Q_i} \int_0^{\infty} e^{-rt} \left( A\phi_i(t) - \frac{B}{2}\phi_i^2(t) - \frac{s}{2}P(t)^2 \right) dt \tag{9.1}$$

subject to

$$\dot{P}(t) = \Sigma_{i=1}^{n} \varepsilon_i(t) - kP(t) \tag{9.2}$$

with

$$P(0) = P_0 \tag{9.3}$$

and where $r$ denotes the discount rate which is assumed to be constant and identical for all countries, and $\phi_i$ denotes the rate of consumption of country $i$. Also, $\varepsilon_i$ denotes country $i$'s instantaneous emissions of pollution generated by the production of its consumption and is given by

$$\varepsilon_i = \theta\phi_i, \tag{9.4}$$

where $\theta$ is an exogenous parameter that captures the ratio of emissions to output. Moreover, $A,\ B$, and $s$ are positive parameters. We note that, in (9.1), the term $\left(A\phi_i(t) - \frac{B}{2}\phi_i^2(t)\right)$ denotes country $i$'s utility from consumption, whereas the term $\frac{s}{2}P(t)^2$ denotes pollution damage faced by each country from the accumulated stock of the pollutant at instant $t$, given by $P(t)$. The stock of pollution, $P(t)$, is increasing in the aggregate emissions, $\Sigma_{i=1}^{n}\varepsilon_i(t)$, and decreasing at the rate $k > 0$, where $k$ denotes the natural rate of decay of the stock, as reflected in (9.2).

The set of strategies considered is the set of stationary Markovian strategies: emissions at each moment depend on the stock of pollution at the moment only. A Markov perfect Nash equilibrium in linear strategies can be characterized (see Dockner & Long 1993 and Dockner et al. 2000).

Such a game admits a unique linear equilibrium and a continuum of equilibria with non-linear strategies (Dockner & Long 1993).

**Proposition 1 (Benchekroun and RayChaudhuri 2014)** *For* $P < \bar{P}(\theta) \equiv \frac{1}{\theta\alpha}(A - \theta\beta)$, *the vector* $(Q, .., Q)$

$$Q_i^*(P;\theta) = Q(P;\theta) \equiv \frac{1}{B}(A - \beta\theta - \alpha\theta P),\ i = 1, \ldots, n \tag{9.5}$$

*constitutes a Markov perfect linear equilibrium and discounted net welfare is given by*

$$W_i(P;\theta) = -\frac{1}{2}\alpha P^2 - \beta P - \mu, \ i = 1, \ldots, n \tag{9.6}$$

*where*

$$\alpha = \frac{\sqrt{B\left(B(2k+r)^2 + (2n-1)4s\theta^2\right)} - (2k+r)B}{2(2n-1)\theta^2}$$

$$\beta = \frac{An\alpha\theta}{B(k+r) + (2n-1)\alpha\theta^2}$$

$$\mu = -\frac{(A-\beta\theta)(A-(2n-1)\beta\theta)}{2Br}.$$

*The steady state level of pollution*

$$P^{SS}(\theta) = \frac{n\theta(A-\theta\beta)}{Bk + n\alpha\theta^2} > 0 \tag{9.7}$$

*is globally asymptotically stable.*

***Proof*** We use the undetermined coefficient technique (see Dockner et al. 2000 Chapter 4) to derive the linear Markov perfect equilibrium. The details are omitted (see Proposition 1 of Dockner and Long (1993) for the case where $\theta = 1$).

We note that when the stock of pollution is beyond a certain threshold $\bar{P}(\theta)$ emissions are zero and that the steady state stock of pollution $P^{SS}(\theta) < \bar{P}(\theta)$ for all $\theta \geq 0$. □

## 9.3 Adoption of a Cleaner Technology

The extent to which a production technology is clean is captured by the emissions to output ratio, $\theta$. The smaller is $\theta$, the cleaner the technology. We assume that as a cleaner technology becomes available, it is immediately adopted by all countries. This allows us to isolate the effect of the technology on the strategic behavior of countries in the mitigation game. We note that other types of strategic behavior may be examined by endogenizing the timing of adoption, which is beyond the scope of this analysis.

It can be shown that the adoption of a cleaner technology has an ambiguous impact on the equilibrium emissions and on the equilibrium long-run stock of pollution (see Propositions 2 and 3 in Benchekroun & RayChaudhuri 2014). In particular, we can have

$$\frac{dP^{SS}}{d\theta} < 0.$$

Moreover, we have

$$E_\theta (P; \theta) \leq (>) 0 \text{ for all } P \geq (<) \tilde{P},$$

where $E(P; \theta)$ denotes the emissions that are associated with the equilibrium production strategy. That is, the adoption of a cleaner technology results in a decrease of emissions in the short run only when the stock of pollution is below a certain level $\tilde{P}$.

This potential discrepancy between long-run and short-run impact of a technological innovation on countries' emissions can only be captured within a dynamic framework. A static framework would provide misleading insights.

The detrimental effect of implementing a cleaner technology on the environment occurs because the cleaner technology reduces the damage from pollution at the margin, providing countries with an incentive to emit more. Since this holds for all $n \geq 1$, it follows that a cleaner technology may result in a larger pollution stock for all $n \geq 1$. The greater the number of countries, the greater the free-riding incentive of each country within this transboundary pollution game. Therefore, when faced with the cleaner technology, each country increases its emissions more the larger is $n$.

This ambiguous impact of a cleaner technology on emissions leads to a follow-up analysis: does a cleaner technology result in an increase in welfare? Indeed, while it may result in larger emissions and damages from pollution, is it possible that the resulting increase in production generates enough gains in utility to compensate the increase in damages from pollution? Benchekroun and RayChaudhuri (2014) show that the answer to this question turns out to be ambiguous as well: a decrease in $\theta$ may result in a decrease in $W_i(P; \theta)$. Using the Hamilton Jacobi Bellman equation associated with a player's problem,

$$rW(P; \theta) = U(Q) - D(P) + W_P(P; \theta)(n\theta Q - kP),$$

along with the envelop theorem, allows us to obtain the following:

$$rW_\theta (P; \theta) = (n-1)\theta W_P Q_\theta + W_{P\theta}(n\theta Q - kP) + nQW_P. \tag{9.8}$$

At $P = P^{SS}(\theta)$, one can show that

$$r\ W_\theta|_{P=P^{SS}(\theta)} = (n-1)\, E_\theta W_P + QW_P. \tag{9.9}$$

Benchekroun and RayChaudhuri (2014) establish that $W_\theta$ may well be positive. This is first done analytically for the case of a marginal decrease in $\theta$ and then numerically for a set of plausible values of the parameters of the model. Note that $W_\theta(P; \theta)$ gives the impact of a (small) change of $\theta$ on welfare at a given stock of pollution. It captures the change in welfare throughout the transition phase from an initial stock $P$ to the steady state.

***Main Proposition (Benchekroun and RayChaudhuri 2014)*** *For any $n > 1$, there exists $\bar{s} > 0$ such that $W_\theta|_{P=P^{SS}(\theta)} > 0$ for all $s > \bar{s}$.*

***Proof*** The formal proof is available in Benchekroun and RayChaudhuri (2014).

Benchekroun and RayChaudhuri (2014) determine the conditions under which implementing a cleaner technology by decreasing $\theta$ decreases welfare. The greater the damage parameter, $s$, the greater the impact on countries' emissions since a cleaner technology leads to greater decrease in pollution damage at the margin. Moreover, the greater is $k$ or $r$, the less important the link between current emissions and the stock of pollution. Hence, countries emit more when $k > \bar{k}$ or $r > \bar{r}$ under the cleaner technology. □

## 9.4 Cleaner Technologies and Climate Change: Application

In this section, we apply our analysis to the specific case of climate change. There is no consensus on the values of the different parameters within this context. Benchekroun and RayChaudhuri (2014) proceed by borrowing values from the existing literature and conduct a sensitivity analysis of the results. The benchmark case has the following parameter values summarized in Table 9.1.

Where $x$ denotes the percentage of world GDP lost due to a change in temperature if the stock of pollution doubles relative to the current level. That is,

$$x(\text{World GDP}) = nD(P),$$

where $D(P)$ refers to pollution damage and is given by $\frac{s}{2}P(t)^2$. This allows us to obtain $s = 2\frac{x(\text{World GDP})}{nP^2}$.[10] Taking into account market and non-market impacts, Heal (2009) estimates that the cost could be 10% of world income. The Stern Review uses 5% as an estimate of $x$. However, with the risk of catastrophe, the 95th percentile is estimated to be 35.2% loss in global per-capita GDP by 2200. Thus,

**Table 9.1** Parameter values

| Parameter | Benchmark case |
|---|---|
| $x$ | 0.025 |
| $n$ | 10 |
| $k$ | 0.005 |
| $r$ | 0.025 |

[10] List and Mason (2001) and Karp and Zhang (2012) have used the same approach to derive the numerical value of the pollution damage parameter. We note that this numerical simulation determines the impact of a reduction in $\theta$ on the present value of GDP net of damages.

2.5% in the benchmark case is a conservative estimate, and results for $x = 5\%$ and $x = 10\%$ are also presented.

Let the relative change in welfare as $\theta$ changes from $\theta_0$ to $\theta$ be given by $G(P, \theta, \theta_0)$ :

$$G(P, \theta, \theta_0) = \frac{W(P; \theta)|_{\theta} - W(P; \theta)|_{\theta=\theta_0}}{W(P; \theta)|_{\theta=\theta_0}}.$$

In the following calculations, we update the values used in Benchekroun and RayChaudhuri (2014). In particular, by updating the values of emissions and GDP, we obtain updated values of $\theta_0$ and $s$. The value of the world GDP in 2022 is \$100,880 billion,[11] total $CO_2$ emissions from fossil fuel combustion and land use change in 2022 is 11.045 GtC,[12] and the short-term decay rate of emissions is 36% (Newell and Pizer (2003).[13] This gives us the value of $\theta$ in 2022, which we use as the value of $\theta_0$ in the following discussion:

$$\theta_0 = \frac{\left(11.045 \times 10^9\right)(3.67)(0.64)}{100,880 \times 10^9} = 2.5716 \times 10^{-4} tCO_2/\$.$$

The updated value of $s$ is given by:

$$s = 2\frac{x(100,880 \times 10^9)}{10\,(590 \times 3.67)^2},$$

where the pre-industrial level of stock of $CO_2$ is given by 590 GtC, and 3.67 represents the conversion rate from units of carbon to units of $CO_2$.

Figure 9.1 shows emissions as a function of stock for two different values of $\theta$. We see that for sufficiently large stocks, emissions associated with the lower $\theta$ are higher.

Figure 9.2 gives a plot of $G\left(P^{SS}(\theta_0), \theta, \theta_0\right)$ where $\theta_0$ is set to $2.5716 \times 10^{-4} tCO_2/\$$ of GDP. The parameter $B$ is set to $\theta_0^2$ so that $\theta = \theta_0$ to retrieve the same specification as Dockner and Long (1993), van der Ploeg and de Zeeuw (1992), List and Mason (2001), and Hoel and Karp (2001). Note that in Fig. 9.2, we have $W(P; \theta)|_{\theta=\theta_0} < 0$, and therefore, when $G(P, \theta, \theta_0) > 0$ we have $W(P; \theta)|_{\theta} - W(P; \theta)|_{\theta=\theta_0} < 0$.

A reduction in emissions per output may result in a decrease in welfare. For a cleaner technology to result in an increase in welfare with respect to the benchmark case $\theta_0$, it has to be "substantially" cleaner than the existing technology. We observe

---

[11] See https://data.worldbank.org/indicator/NY.GDP.MKTP.CD.

[12] See https://www.carbonbrief.org/analysis-global-co2-emissions-from-fossil-fuels-hit-record-high-in-2022/.

[13] i.e. 64% of emissions adds to the stock in any given year). Also, 3.67 represents the conversion rate from units of carbon to units of $CO_2$.

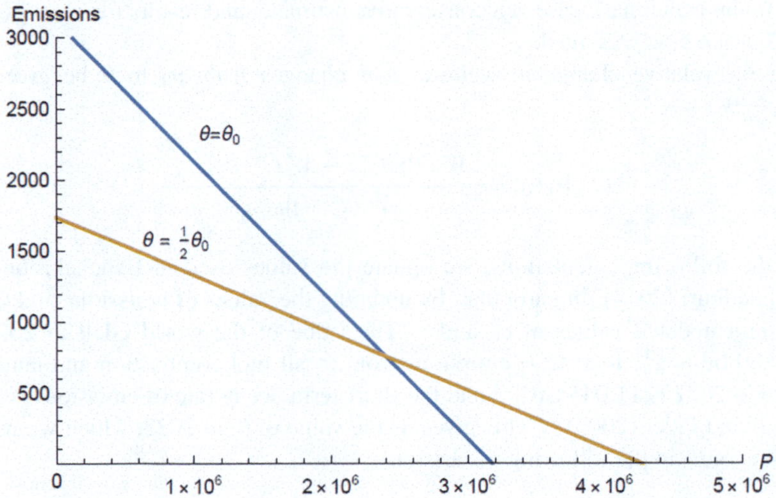


**Fig. 9.1** Emissions as a function of $P$ as $\theta$ changes

that $\theta$ must fall below $\tilde{\theta}_0 = 0.6478 \times 10^{-4} tCO_2/\$$ or 74.8% drop in $\theta$. The "required" drop in $\theta$ to start observing an increase in welfare is even larger for larger values of the damage from pollution: the threshold $\tilde{\theta}_0$ falls to $0.381 \times 10^{-4} tCO_2/\$$ (i.e., a decrease of 85.2%) when we use $x = 5\%$, and to $0.2338 \times 10^{-5} tCO_2/\$$ (i.e., a decrease of 90.9%) when $x = 10\%$.

Benchekroun and RayChaudhuri (2014) present a similar analysis with similar qualitative findings to that illustrated in Fig. 9.2. While Fig. 9.2 shows that scenario for 2022, Benchekroun and RayChaudhuri (2014) show the scenario for 2008. Compared to 2008, we note that $\theta_0$ is lower ($\theta_0$ was set to $3.8315 \times 10^{-4} tCO_2/\$$ for 2008 as compared to $2.5716 \times 10^{-4} tCO_2/\$$ for 2022). However, as is clear from Fig. 9.2 and from the ensuing discussion, the problem still persists such that small or intermediate decreases in $\theta$ starting from the 2022 level of $\theta_0$ are expected to lead to decreases in welfare.

Figure 9.3 gives a plot of $W_\theta|_{P=P^{SS}(\theta_0)}$ as a function of $x$.

For the benchmark case, for $x > 0.6\%$ we have $W_\theta|_{P=P^{SS}(\theta_0)} > 0$. A marginal decrease in emissions per output ratio reduces welfare. The relationship of $W_\theta|_{P=P^{SS}(\theta_0)}$ with respect to $x$ (which is a proxy for $s$) mirrors the result obtained analytically for the behavior of $W_\theta|_{P=P^{SS}(\theta)}$ in the limit case where $s \to \infty$. The larger the damage parameter, the more likely a decrease of the emissions per output ratio will be welfare reducing.

Let $Z \equiv \frac{P_0}{P^{SS}(\theta_0)}$. That is, $Z$ is a parameter that sets the initial level of the stock of pollution relative to the steady state stock of pollution. Figure 9.4 shows that the graph of $W_\theta|_{P=Z*P^{SS}(\theta_0)}$ is a strictly increasing function of $Z$.

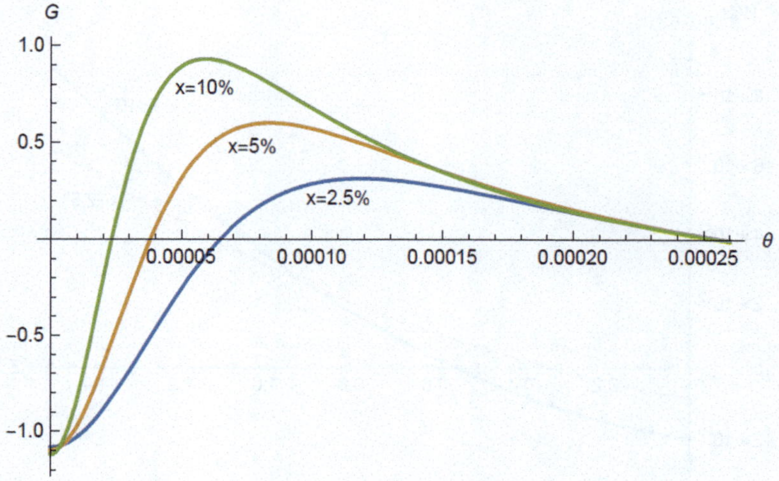


**Fig. 9.2** $G$ at $P = P^{SS}(\theta_0)$ as a function of $\theta$

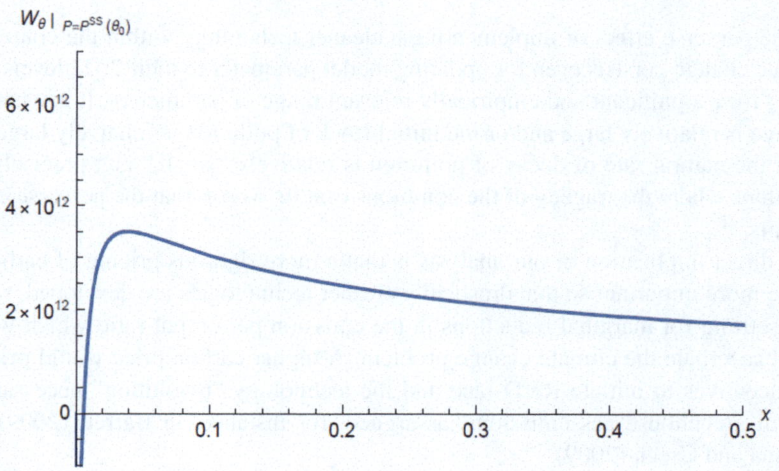


**Fig. 9.3** $W_\theta|_{P=P^{SS}(\theta_0)}$ as a function of $x$

Figure 9.4 shows that $W_\theta|_{P=Z*P^{SS}(\theta_0)}$ is positive for $Z > \tilde{Z} = 0.549$. The larger the stock of pollution at which we introduce a cleaner technology, the more likely this will result in a welfare loss. The value of $\tilde{Z}$ decreases to 0.325 when $x = 10\%$.

In a world where countries continue to set their emission levels non-cooperatively, incremental innovations that result in small reductions in the emission per output ratio can actually be harmful instead of helpful. In fact, they are most likely to reduce welfare in the most dire circumstances, i.e., when the damage is important and/or the stock of pollution is large enough, and nature is least able to absorb pollution.

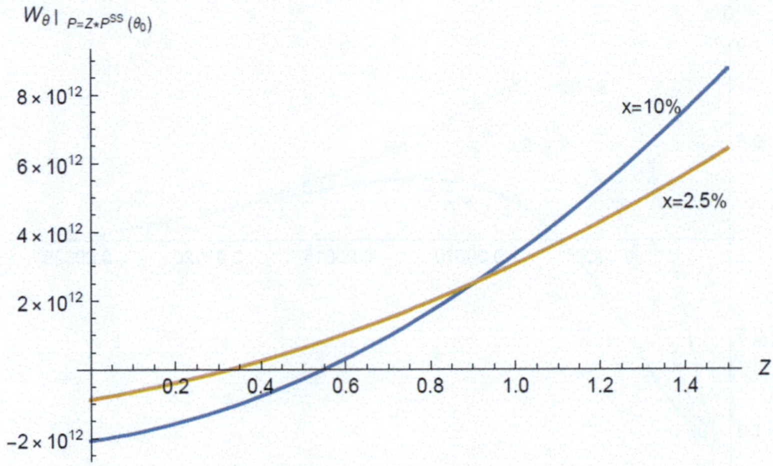


**Fig. 9.4** $W_\theta|_{P=Z*P^{SS}(\theta_0)}$ as a function of $Z$

The perverse effect of implementing a cleaner technology within the context of climate change persists even for updating model parameter to their 2022 levels: it is strong for a significant and empirically relevant range of parameters. It is when the damage is relatively large and/or the initial stock of pollution is relatively large and when the natural rate of decay of pollution is relatively "small," i.e., precisely the situations where the tragedy of the commons is at its worse, that the perverse effect prevails.

A direct implication of our analysis is that a more rigorous pricing of carbon is all the more important so that drastically cleaner technologies are developed, rather than settling for marginal reductions in the emission per output ratio which would only exacerbate the climate change problem. A higher carbon price would provide the incentives to initiate R&D race and the technology "revolution" necessary to control greenhouse gas emissions, as argued, for instance, in Barrett (2009) and Galiana and Green (2009).

## 9.5 Concluding Remarks

Technological innovations that result in a cleaner production process can have an ambiguous impact on emissions; the short-run impact on emissions depends on the initial stock pollution and can be inversely related to the long-run impact. This underscores the necessity of using a fully fledged dynamic framework to assess the impact of a technological innovation since a static framework would not be able capture this ambiguity.

Innovations that marginally reduce the emission per output ratio can backfire and result in an increase of pollution and even welfare. This occurs for "realistic" values of the parameters of the model. Our finding that emissions per output ratio and world emissions can evolve in opposite directions is supported by recent anecdotal evidence within the context of climate change. While the world's emissions per output ratio decreased from $3.8315 \times 10^{-4} tCO_2/\$$ in 2008 to $2.5716 \times 10^{-4} tCO_2/\$$ in 2022, world's emissions of $CO_2$ have continued to steadily increase.

This is because technological innovations that marginally reduce the emission per output ration have two effects. First, the direct effect is to decrease emissions if the quantity produced by each player remains unchanged. Second, the indirect effect is to increase emissions since quantity produced increases as emissions become less damaging at the margin. The latter effect can outweigh the former.

There are two main policy recommendations that can be drawn from our analysis. First, investments in technological innovations should not be seen as substitutes to the urgent need to an ambitious multilateral effort to mitigate emissions: these two objectives should be pursued jointly. Second, policies aimed at fostering research and development and innovations to tackle climate change should focus on very ambitious innovations and research and development (R&D) projects instead of supporting a wider range of potential projects but with relatively smaller improvement over existing technologies (see e.g., Matos et al. (2022) and Pooler, 2021). These policies include providing substantial subsidies and tax credits for R&D.

A promising line of future research that would contribute to gain insights into the last policy recommendation is to enrich our model to allow for an endogenous technological innovation that reduces the emission per output ratio: a model that embeds this framework and where investment in R&D to reduce emissions per output is taken into account. The adverse impact of clean technologies would not take place in the presence of a well-designed international limit over emissions. Although this is intuitive, the impact of quotas in dynamic games is complex (see, e.g., Dockner & Haug 1990, 1991), as is the impact of cleaner technologies on the size of stable international environmental agreements and the level of emissions control that can be self-enforced in such agreements (see Benchekroun & RayChaudhuri 2015). These issues warrant a closer examination.

**Acknowledgments** We thank the Canadian Social Sciences and Humanities Research Council (SSHRC) for financial support.

## References

Barrett, S. (2009). The coming global climate-technology revolution. *Journal of Economic Perspectives, 23*, 53–75.

Benchekroun, H., & RayChaudhuri, A. (2014). Transboundary pollution and clean technologies. *Resource and Energy Economics, 36*(2), 601–619.

Benchekroun, H., & RayChaudhuri, A. (2015). Cleaner technologies and the stability of international environmental agreements. *Journal of Public Economic Theory, 17*(6), 887–915.

Chenavaz, R. Y., Dimitrov, S., & Figge, F. (2021). When does eco-efficiency rebound or backfire? An analytical model. *European Journal of Operational Research, 290*, 687–700.
Dockner, E., & Haug, A. A. (1990). Tariffs and quotas under dynamic duopolistic competition. *Journal of International Economics, 29*, 147–59.
Dockner, E., & Haug, A. A. (1991). Closed-loop motive for voluntary export restraints. *Canadian Journal of Economics, 24*, 679–685.
Dockner, E. J., Jorgensen, S., Long, N. V., & Sorger, G. (2000). *Differential games in economics and management science*. Cambridge University Press.
Dockner, E. J., & Long, N. V. (1993). International pollution control: Cooperative versus noncooperative strategies. *Journal of Environmental Economics and Management, 24*, 13–29.
Dutta, P. K., & Radner, R. (2006). Population growth and technological change in a global warming model. *Economic Theory, 29*, 251–270.
Galiana, I., & Green, C. (2009). Let the technology race begin. *Nature, 462*, 570–571.
Gerlagh, R. (2011). Too much oil. *CESifo Economic Studies, 57*, 79–102.
Greening, L. A., Greene, D. L., & Difiglioc, C. (2000). Energy efficiency and consumption — the rebound effect — a survey. *Energy Policy, 28*, 389–401.
Heal, G. (2009). Climate economics: A meta-review and some suggestions for future research. *Review of Environmental Economics and Policy, 3*, 4–21.
Hoel, M. (2011). The green paradox and greenhouse gas reducing investments. *International Review of Environmental and Resource Economics, 5*, 353–379.
Hoel, M., & Karp, L. (2001). Taxes and quotas for a stock pollutant with multiplicative uncertainty. *Journal of Public Economics, 82*, 91–114.
Jørgensen, S., & Zaccour, G. (2001). Time consistent side payments in a dynamic game of downstream pollution. *Journal of Economic Dynamics and Control, 25*(12), 1973–1987.
Jørgensen, S., Martín-Herrán, G., Zaccour, G. (2010) Dynamic games in the economics and management of pollution, *Environmental Modeling & Assessment 15*, 433–467.
Karp, L., & Zhang, J. (2012). Taxes versus quantities for a stock pollutant with endogenous abatement costs and asymmetric information. *Economic Theory, 49*, 371–409.
Langinier, C., & RayChaudhuri, A. (2020). Green technology and patents in the presence of green consumers. *Journal of the Association of Environmental and Resource Economists, 7*(1), 73–101.
List, J., & Mason, C. (2001). Optimal institutional arrangements for transboundary pollutants in a second-best world: Evidence from a differential game with asymmetric players. *Journal of Environmental Economics and Management, 42*, 277–296.
Matos, S., Viardot, E., Sovacool, B. K., Geels, F. W., & Xiong, Y. (2022). Innovation and climate change: A review and introduction to the special issue. *Technovation*, 102612.
Newell, R. G., & Pizer, W. A. (2003). Regulating stock externalities under uncertainty. *Journal of Environmental Economics and Management, 45*, 416–432.
Pooler, M. (2021). 'Green Steel': The Race to Clean up One of the World's Dirtiest Industries. Financial Times. Feb 15. https://www.ft.com/content/46d4727c-761d-43ee-8084-ee46edba491a.
Quentin Grafton, R., Kompas, T., & Long, N. V. (2012). Substitution between biofuels and fossil fuels: Is there a green paradox? *Journal of Environmental Economics and Management, 64*, 328–341.
Sinn, H. W. (2012). *The green paradox: A supply-side approach to global warming*. MIT Press.
Sorrell, S., & Dimitropoulos, J. (2008). The rebound effect: Microeconomic definitions, limitations and extensions. *Ecological Economics, 65*, 636–649.
van der Ploeg, F., & de Zeeuw, A. (1992). International aspects of pollution control. *Environmental and Resource Economics, 2*, 117–139.
van der Ploeg, F., & Withagen, C. (2012). Is there really a Green Paradox? *Journal of Environmental Economics and Management, 64*, 342–363.

# Chapter 10
# On the Effects of an Increase in the Number of Abaters in Pollution Abatement Games

**Luca Colombo and Paola Labrecciosa**

**Abstract** We study the effects of an increase in the number of abaters in pollution abatement games, first in a static, then in a dynamic (continuous-time) game. In both games, we assume that $m$ countries/agents agree on taking action to reduce the stock of pollution, which is a public bad, whereas $n - m$ countries free ride on the abatement levels of the abaters. Moreover, we assume that abaters can either coordinate on their contributions or not. In the static game, both in the coordination and the non-coordination scenario, an increase in $m$ leads to a decrease in the stock of pollution and to an increase in social welfare. In the dynamic game, instead, both in the coordination and the non-coordination scenario, an increase in $m$ may result in a higher steady-state stock of pollution and a lower social welfare, depending on the "business-as-usual" level of output.

**Keywords** Differential games · Pollution abatement · Mode of cooperation

**JEL Classification** Q2, Q52, C73

---

Part of this chapter is based on the discussion paper "A Dynamic Analysis of Climate Change Mitigation with Endogenous Number of Contributors: Loose vs Tight Cooperation" by Luca Colombo, Paola Labrecciosa, and Ngo Van Long, 2019, available at http://hdl.handle.net/10086/30916.

---

L. Colombo (✉)
Rennes School of Business, Rennes, France
e-mail: luca.colombo@rennes-sb.com

P. Labrecciosa
ESSCA School of Management (Paris Campus), Boulogne-Billancourt, France
e-mail: paola.labrecciosa@essca.fr

 
F. El Ouardighi, G. Feichtinger (eds.), *The Unaffordable Price of Static Decision-making Models*, International Series in Operations Research & Management Science 365, https://doi.org/10.1007/978-3-031-88638-6_10

## 10.1 Introduction

Many global public goods such as widespread peace, financial stability, public health, and climate change mitigation are funded predominantly through voluntary contributions.[1] As argued in Colombo et al. (2022), the attainment of an efficient outcome in the presence of global public goods requires international cooperation. However, experience from climate change policy indicates that full cooperation among all countries is difficult to achieve. Partial cooperation seems to be a more realistic prospect due to conflicting national interests, disagreements on what constitutes a fair burden, and a general distrust among countries.

In this chapter, we focus on climate change mitigation and consider two modes of cooperation, which we refer to as tight and loose cooperation (see Colombo et al., 2019). The former corresponds to the traditional case of full cooperation in which participating countries choose their abatement levels with the aim of maximizing the sum of payoffs of all coalition members and the latter to the perhaps more realistic case in which participating countries agree to contribute to the abatement of the stock of pollution but choose their abatement levels with the aim of maximizing their own payoff exclusively. In the tight mode of cooperation, countries coordinate on their emission levels, while in the loose mode of cooperation, they do not. In this respect, the tight mode of cooperation can be thought of as a centralized, top-down approach (e.g., the Kyoto Protocol), while the loose mode of cooperation can be thought of as a decentralized, bottom-up approach (e.g., the Paris Agreement). In our theoretical investigations, irrespective of the mode of cooperation, we assume that the population of countries/agents is divided into two groups: a group of participating countries, each making a positive contribution toward the abatement of the stock of pollution, and a group of nonparticipating countries, whose contribution is nil. Given the public good nature of climate change mitigation, these countries free ride on the contributions made by the participating countries. Unlike Colombo et al. (2022) and many other papers in which the contribution stage is preceded by a participation stage (see, e.g., Carraro & Siniscalco, 1993; Barrett, 1994; Rubio & Ulph, 2006; Eichner & Pethig, 2013), we restrict attention to the contribution stage and assume that the size of the two groups is exogenously given.[2] Another important difference with respect to Colombo et al. (2022) is that instead of assuming linear benefits and costs, we use quadratic functions for both benefits and costs, with

[1] Classical references on the voluntary provision of public goods include Chamberlin (1974), Bergstrom et al. (1986), Bergstrom et al. (1986), Cornes and Sandler (1986), and Andreoni (1988). On the voluntary provision of public goods in dynamic settings, see Fershtman and Nitzan (1991), Varian (1994), Wirl (1996), Marx and Matthews (2000), Itaya and Shimomura (2001), Yanase (2006), Long and Shimomura (2007), Benchekroun and Long (2008), Fujiwara and Matsueda (2009), Battaglini et al. (2014), Georgiadis (2015, 2017), and Bowen et al. (2019), inter alia.

[2] For the participation stage, the usual stability concept is that dating back to d'Aspremont et al. (1983). On the stability of coalitions see, e.g., Rubio and Ulph (2007) and de Zeeuw (2008), who build on earlier works by Carraro and Siniscalco (1993) and Barrett (1994).

benefits derived in terms of the difference between the "business-as-usual" level of output and the abatement level.

There are many examples of successful loose cooperation in world history. As to climate change policy, consider the Paris International COP21 Conference on Climate Change. Countries have agreed on an overall objective of limiting global warming to 2 degrees C relative to the pre-industrial temperature, but no country is required to set a specific target by a specific date. Signatories are free to determine their own target, and there is no penalty if a target is not met. Outside the domain of environmental economics, of note is the Hanseatic League, a Central European loose confederation of merchant guilds and market cities that came to dominate Baltic maritime trade for over 300 years, reaching its peak in the fifteenth century, with over 100 member cities (see Atatüre, 2008). A more recent example is the Association of Southeast Asian Nations (ASEAN) which promote free trade among member countries without requiring them to abide by strict rules. In modern democracies, political parties are notable instances of loose cooperation. In the USA, for example, Republicans and Democrats can move in and out of their parties without penalties, and while political donations are encouraged, party members are not asked to commit to specific donations. As argued in Colombo et al. (2019), an important feature of loose cooperation is that it allows individual countries/agents the flexibility to respond to idiosyncratic shocks. Many of these shocks are related to political economy considerations, such as discontents from a powerful segment of the electorate. Shocks of this nature are quite often private information (i.e., not verifiable by third parties) and therefore state-contingent transfer payments are not feasible (Bagwell & Staiger, 2005; Amador & Bagwell, 2013). These are important considerations. However, for simplicity, as a first step, we do not model idiosyncratic shocks and asymmetry among countries/agents (apart from the distinction between abaters and free riders).

The main purpose of our analysis is to study the impact of an increase in the number of abaters, either coordinating or not, on the stock of pollution and social welfare. First, we consider a static model in which, by definition, time does not play any role. Second, we extend the analysis to an infinite-horizon model where time is continuous. Both in the static and the dynamic model, abaters act either non-cooperatively or cooperatively, while free riders remain passive players. In most of the analysis, for analytical tractability, we assume that there are only two countries/agents, and focus on the effects of an increase in the number of abaters from 1 to 2. While in the static model (both in the non-coordinating and the coordinating scenario) an increase in the number of abaters unambiguously leads to a decrease in the stock of pollution and to an increase in social welfare, in the dynamic model (both in the non-coordination and the coordination scenario) whether the traditional static result arises crucially depends on the "business-as-usual" level of output. Counterintuitively, conditions exist under which increasing the number of abaters is detrimental to the environment and social welfare. The main implication of this finding for climate change policy is that increased participation in international environmental agreements (IEAs) is not necessarily socially desirable, irrespective of the mode of cooperation. While it is typically argued in the static literature on IEAs that larger coalitions are welfare-superior to smaller ones, our analysis suggests that this might not be the case in a dynamic setting. The price of not adequately considering the time dimension might be a wrong policy recommendation.

The remainder of this paper is organized as follows. The static model is specified in Sect. 10.2, which also contains the analysis of the non-coordination and the coordination scenarios. The dynamic model and the analysis of the two modes of cooperation are provided in Sect. 10.3. Section 10.4 evaluates the impact of an increase in the number of contributors on the stock of pollution and social welfare. Section 10.5 concludes.

## 10.2 The Static Game

Consider an economy populated by $n \geq 2$ countries/agents divided into two groups: a group $G^C$ consisting of $m$ contributors to the abatement of a public bad, and a group $G^N$ consisting of $n - m$ free riders whose contribution is nil. Contributors are denoted by the index $i = 1, ..., m$, where $m \leq n$, and free riders by the index $j$. The size of the groups is exogenously given. Each agent's maximum productive capacity of the final consumption good is a positive number $a$, which we refer to as their "business-as-usual" level of output. Each unit of output generates one unit of emission of a pollutant. The stock of pollution is a public bad. Individuals realize that if they produce the consumption good at their maximum productive capacity, they will each add $a$ units of pollutant to the stock of pollution. Cutting output below the maximum capacity is referred to as "abatement." If an individual $i$ chooses the abatement level $x_i$, where $x_i \in [0, a]$, she/he will have only $a - x_i$ units of output to consume. We assume that the direct utility derived from consumption is quadratic:

$$U(a - x_i) \equiv (a - x_i) - (a - x_i)^2, \tag{10.1}$$

where $a < 1/2$ so that any $x_i \in [0, a]$ constitutes a sacrifice of direct utility. The ex-ante stock of pollution is $k_0 \geq 0$. The ex-post stock of pollution is defined as

$$k \equiv k_0 + na - X, \tag{10.2}$$

where $X \in [0, ma]$ denotes the sum of contributions. The *ex-post* stock of pollution inflicts a damage $bk^2$ to each agent, where $b > 0$. From (10.1) and (10.2), the (net) utility function of a contributor is given by[3]

[3] We will impose parameter restrictions which guarantee that $u_i$ is decreasing in $k$ and increasing in $a$ and that consumption net of the abatement, $a - x_i$, is nonnegative.

$$u_i(k_0, x_i) \equiv U(a - x_i) - bk^2 \equiv (a - x_i) - (a - x_i)^2 - b\,(k_0 + na - X_{-i} - x_i)^2, \tag{10.3}$$

where $X_{-i} \in [0, (m - 1)a]$ denotes the sum of contributions by all contributors except $i$.

For future reference, we provide the definition of social welfare:

$$W \equiv mu_i(k_0, x_i) + (n - m)\,u_j(k_0), \tag{10.4}$$

where the utility function of a free rider is given by

$$u_j(k_0) \equiv a - a^2 - b\,(k_0 + na - X)^2.$$

### 10.2.1 Non-coordination Scenario

In the non-coordination scenario, participating countries agree to contribute to the abatement of the stock of pollution but are left free to choose their levels of contribution. Contributor $i \in G^C$ chooses $x_i$ independently with the aim of maximizing $u_i(k_0, x_i)$ given in (10.3). The first-order necessary and sufficient condition for utility maximization is given by

$$-1 + 2(a - x_i) + 2b(k_0 + na - X_{-i} - x_i) = 0,$$

from which we can derive the best-response function of a generic contributor:

$$x_i\,(X_{-i}) = \frac{2\,[a + b\,(k_0 + an - X_{-i})] - 1}{2\,(1 + b)}.$$

Clearly, contributions are strategic substitutes: an increase in $X_{-i}$ leads to a decrease in $x_i\,(X_{-i})$. Let $x_i^*$ denote the symmetric Nash equilibrium level of individual contributions. It follows that

$$-1 + 2(a - x_i^*) + 2b(k_0 + na - mx_i^*) = 0.$$

**Proposition 10.1** *The Nash equilibrium level of individual contributions is equal to*

$$x_i^* = \frac{2b(k_0 + na) + 2a - 1}{2\,(1 + bm)},$$

*which is positive if $a > \widetilde{a}$, with*

$$\widetilde{a} = \frac{1 - 2bk_0}{2\,(1 + bn)}.$$

### 10.2.2 Coordination Scenario

In the coordination scenario, the level of contribution by country/agent $i \in G^C$ is chosen by the coalition with the aim of maximizing the sum of utilities of all participating countries, i.e., $\sum_{i=1}^{m} u_i(k_0, x_i)$. The first-order necessary and sufficient condition for joint utility maximization is given by

$$-1 + 2(a - x_i^{**}) + 2bm(k_0 + na - mx_i^{**}) = 0,$$

where $x_i^{**}$ denotes the level of individual contributions under coordination.

**Proposition 10.2** *The level of individual contributions under coordination is equal to*

$$x_i^{**} = \frac{2\left[bk_0 m + a\left(1 + bmn\right)\right] - 1}{2\left(1 + bm^2\right)},$$

*which is positive if* $a > \tilde{a}^C$, *with*

$$\tilde{a}^C = \frac{1 - 2bk_0 m}{2\left(1 + bmn\right)} \leq \tilde{a}.$$

## 10.3 The Dynamic Game

We extend the analysis of the previous section to a dynamic setting. Time is continuous and denoted by $t \in [0, \infty)$. There are $n \geq 2$ infinitely lived agents. As in the previous section, agents are divided into two groups, namely, a group $G^C$ consisting of $m$ active players (each contributing to the abatement of a public bad) and a group $G^N$ consisting of $n-m$ free riders, whose contribution is nil. The size of each group is exogenously given and constant over time. Contributors are denoted by the index $i = 1, ..., m$, where $m \leq n$, and free riders are denoted by the index $j$. The instantaneous direct utility derived from consumption is quadratic:

$$U(a - x_i(t)) \equiv [a - x_i(t)] - [a - x_i(t)]^2, \tag{10.5}$$

where $a < 1/2$. (10.5) is the dynamic counterpart of (10.1). The stock of pollution at $t$ is denoted by $k(t) \geq 0$. At any $t \geq 0$, the stock $k(t)$ inflicts a damage flow

$bk(t)^2$ to each agent, where $b > 0$. The instantaneous (net) utility function of a contributor is defined as[4]

$$u_i(k(t), x_i(t)) \equiv U(a - x_i(t)) - b[k(t)]^2$$
$$\equiv [a - x_i(t)] - [a - x_i(t)]^2 - b[k(t)]^2, \quad (10.6)$$

where $x_i(t) \in [0, a]$. (10.6) is the dynamic counterpart of (10.3).

The stock of pollution $k(t)$ is assumed to evolve over time according to the following differential equation:

$$\frac{dk(t)}{dt} = na - \sum_{i=1}^{m} x_i(t) - \delta k(t), \quad (10.7)$$

where $\delta$ is the natural decay rate of the stock of pollution, with $\delta > 0$. By (10.7), the addition to the stock of pollution is increasing in $na$ and decreasing in the sum of abatement levels ($\sum_{i=1}^{m} x_i(t)$).

Let $r$ be the positive rate at which future payoffs are discounted. As in the previous section, before analyzing the non-coordination and the coordination scenarios, we provide the definition of social welfare:

$$w(t) \equiv mu_i(k(t), x_i(t)) + (n - m)u_j(k(t)), \quad (10.8)$$

where

$$u_j(k(t)) \equiv a - a^2 - b[k(t)]^2.$$

### 10.3.1 Non-coordination Scenario

The objective functional of contributor $i \in G^C$ is given by

$$J_i \equiv \int_0^{\infty} e^{-rt} \left\{-b[k(t)]^2 + a - x_i(t) - [a - x_i(t)]^2\right\} dt.$$

We assume that contributors use (stationary) feedback strategies, i.e., they condition their contributions at time $t$ on the current level of the stock of pollution, exclusively. Each contributor $i$ takes the strategies of other contributors as given. We must then

[4] As in the previous section, we will impose parameter restrictions which guarantee that $u_i$ is decreasing in $k$ and increasing in $a$, and that consumption net of the abatement, $a - x_i$, is nonnegative.

solve for a non-cooperative feedback equilibrium of the dynamic game. Given a number of contributors $m \geq 1$, non-cooperative feedback equilibrium strategies of a generic contributor must satisfy the following Hamiltonian-Jacobian-Bellman (HJB) equations, where $V_i(k; m)$ denotes the value function of a contributor, given that there are $m$ contributors:

$$r V_i(k; m) = \max_{x_i \in [0,a]} \left\{ -bk^2 + a - x_i - (a - x_i)^2 + \frac{dV_i(k; m)}{dk} (na - x_i - X_{-i}(k) - \delta k) \right\}, \tag{10.9}$$

where $X_{-i}(k) = \sum_{h \neq i}^{m} x_h(k)$. Maximization of the right-hand side of (10.9) gives (for an interior solution):[5]

$$x_i^*(k) = a - \frac{1}{2} \left[ 1 + \frac{dV_i(k; m)}{dk} \right]. \tag{10.10}$$

By inserting (10.10) into (10.9) and imposing symmetry we obtain

$$r V_i(k; m) = \frac{1}{4} \left\{ 1 - 4bk^2 - \left[ \frac{dV_i(k; m)}{dk} \right]^2 \right\} + \frac{dV_i(k; m)}{dk} \left\{ an + \frac{m}{2} \left[ 1 - 2a + \frac{dV_i(k; m)}{dk} \right] - \delta k \right\}. \tag{10.11}$$

Given the linear quadratic structure of the game at hand we guess a value function of the form

$$V_i(k; m) = A\frac{k^2}{2} + Bk + C, \tag{10.12}$$

where $A$, $B$, and $C$ are constants to be determined. Let $\Delta(m) = \sqrt{4b(2m-1) + (2\delta + r)^2}$. It can be checked that (10.12) with

$$A = \frac{2\delta + r - \Delta(m)}{2m - 1} < 0, \tag{10.13}$$

$$B = \frac{A(2a(n - m) + m)}{A(1 - 2m) + 2(\delta + r)} < 0, \tag{10.14}$$

[5] For an interior solution $x_i^*(k) \in (0, a)$ it must be that $dV_i^C(k)/dk \in (-1, 2a - 1)$. We will verify later that conditions on the parameters of the model exist such that $dV_i^C(k)/dk \in (-1, 2a - 1)$ at any point in time, implying that the equilibrium trajectory of $x_i^*$ remains between 0 and $a$ throughout the entire planning horizon.

and

$$C = \frac{B\,[2m(1-2a) + 4an + B(2m-1)] + 1}{4r},$$

satisfies (10.11) for any $k \in (\underline{k}, \overline{k})$, with $\underline{k} = (2a - 1 - B)/A$ and $\overline{k} = -(1 + B)/A$. For $k \notin (\underline{k}, \overline{k})$ we have a corner solution, either $x_i^* = 0$ or $x_i^* = a$. Specifically, $x_i^* = 0$ for $k < \underline{k}$ and $x_i^* = a$ for $k > \overline{k}$. Note that $\underline{k} \leq 0$ if $a$ is sufficiently close to $1/2$; in that case $x_i^*(k) > 0$ for all $k > 0$.

The above discussion leads to the following proposition.

**Proposition 10.3** *The feedback equilibrium strategy of a contributor is given by* $(i = 1, 2, ..., m)$

$$x_i^*(k) = a - \frac{1 + Ak + B}{2}$$

*for k such that* $x_i^*(k) \in (0, a)$, *where A and B are constants given in (10.13) and (10.14), respectively.*

Two remarks are in order. First, the equilibrium strategy given in Proposition 10.2 is for an interior solution, $x_i^*(k) \in (0, a)$. Given the purpose of our analysis, corner solutions are not interesting: when $x_i^*(k) = 0$ the distinction between contributors and free riders vanishes; when $x_i^*(k) = a$ abatement levels become state-independent and private consumption becomes nil. Second, the equilibrium strategy given in Proposition 10.2 is linear in $k$, with a positive slope equal to $-A/2 > 0$. Thus, $x_i^*(k)$ is increasing in $k$: the higher the stock of pollution, the higher the contribution to the abatement by agent $i = 1, 2, ..., m$, for any given $m \geq 1$.[6] The fact that $x_i^*(k)$ is increasing in $k$ implies that there exists intertemporal strategic substitutability: an increase in $X_{-i}$ leads to a decrease in $k$ which in turn leads to a decrease in $x_i$.

The equilibrium trajectory of the stock of pollution, $k^*(t)$, is the solution to the following first-order differential equation:

$$\frac{dk(t)}{dt} = an - mx_i^*(k(t)) - \delta k(t),$$

with $x_i^*(k)$ given in Proposition 10.2. It can be checked that

[6] As is well-known in the differential game literature, the linear feedback strategy is only one of the infinitely many feedback strategies that satisfy the differential equation resulting from differentiating the maximized HJB equation w.r.t. the state variable. However, value functions associated with nonlinear feedback strategies can be obtained only implicitly, whereas value functions associated with linear feedback strategies are polynomials of degree two and can be easily used for the derivation of the equilibrium strategies.

$$k^*(t) = k^*_{ss} + e^{-\phi t}(k_0 - k^*_{ss}),$$

where

$$\phi = \frac{2\delta(1-m) - m[\Delta(m) - r]}{2(1-2m)} > 0 \tag{10.15}$$

is the speed of convergence to $k^*_{ss}$, with

$$k^*_{ss} = \frac{m(1+B) + 2a(n-m)}{2\delta - Am}. \tag{10.16}$$

It is immediate to verify that $\lim_{t\to\infty} k^*(t) = k^*_{ss}$. The following corollary can then be established.

**Corollary 10.1** *The vector of strategies* $(x^*_1(k), x^*_2(k), ..., x^*_m(k))$ *induces a trajectory of k given by*

$$k^*(t) = k^*_{ss} + e^{-\phi t}(k_0 - k^*_{ss})$$

*where* $\phi > 0$ *is the speed of convergence given in (10.15) and* $k^*_{ss}$ *is the (stable) steady state of k given in (10.16).*

### 10.3.2 Coordination Scenario

We now turn our attention to the coordination scenario. We need to derive $V_i^C(k_0; m)$ under the assumption that those who contribute act cooperatively, i.e., they must coordinate on their contributions. Under cooperation among contributors, the objective functional of contributor $i$ is given by

$$J_i^C \equiv \int_0^\infty e^{-rt}\left\{-b[k(t)]^2 + \left[a - \frac{X(t)}{m}\right] - \left[a - \frac{X(t)}{m}\right]^2\right\} dt,$$

where $X(t)/m$ denotes the coordinated contribution of the representative contributor $i$. Given a number of contributors $m \geq 1$, equilibrium strategies of a generic contributor under cooperative behavior must satisfy the following HJB equations:

$$rV_i^C(k; m) = \max_{X\in[0,na]}\left\{-bk^2 + a - \frac{X}{m} - \left(a - \frac{X}{m}\right)^2 + \frac{dV_i^C(k; m)}{dk}(na - X - \delta k)\right\}. \tag{10.17}$$

Maximization of the right-hand side of (10.17) gives (for an interior solution):[7]

$$X = m\left[a - \frac{1}{2}\left(1 + m\frac{dV_i^C(k;m)}{dk}\right)\right]. \tag{10.18}$$

Inserting (10.18) into (10.17) yields

$$rV_i^C(k;m) = \frac{1}{4}\left\{1 - 4bk^2 + \frac{dV_i^C(k;m)}{dk}\right.$$
$$\left.\times\left[2m(1-2a) - 4k\delta + 4an + m^2\frac{dV_i^C(k;m)}{dk}\right]\right\}. \tag{10.19}$$

As in the non-coordination case, we guess a value function of the form

$$V_i^C(k;m) = A^C\frac{k^2}{2} + B^C k + C^C, \tag{10.20}$$

where $A^C$, $B^C$, and $C^C$ are constants to be determined. Let $\Gamma(m) = \sqrt{4bm^2 + (2\delta + r)^2}$. It can be checked that (10.20) with

$$A^C = \frac{2\delta + r - \Gamma(m)}{m^2} < 0, \tag{10.21}$$

$$B^C = \frac{\alpha\left[2a(m-n) - m\right]}{\alpha m^2 - 2(\delta + r)} < 0, \tag{10.22}$$

and

$$C^C = \frac{1 + B^C m\left(2 - 4a + B^C m\right) + 4aB^C n}{4r}$$

satisfies (10.19) for any $k \in (\underline{\underline{k}}, \overline{\overline{k}})$, with $\underline{\underline{k}} = ((2a-1)/m - B^C)/A^C$ and $\overline{\overline{k}} = -(1/m + B^C)/A^C$. For $k \notin (\underline{\underline{k}}, \overline{\overline{k}})$ we have a corner solution, either $x_i^{**} = 0$ or $x_i^{**} = a$. Specifically, $x_i^{**} = 0$ for $k < \underline{\underline{k}}$ and $x_i^{**} = a$ for $k > \overline{\overline{k}}$. Note that $\underline{\underline{k}} \leq 0$ if $a$ is sufficiently close to 1/2; in that case $x_i^{**}(k) > 0$ for all $k > 0$. The above discussion leads to the following proposition.

[7] For an interior solution $x_i^{**}(k) \in (0, a)$ it must be that $dV_i^C(k; m|coop)/dk \in (-1/m, (2a - 1)/m)$. We will verify later that conditions on the parameters of the model exist such that $dV_i^C(k; m|coop)/dk \in (-1/m, (2a-1)/m)$ at any point in time implying that the equilibrium trajectory of $x_i^{**}$ remains between 0 and $a$ throughout the entire planning horizon.

**Proposition 10.4** *The feedback strategy of a contributor under coordination is given by ($i = 1, 2, ..., m$)*

$$x_i^{**}(k) = a - \frac{1 + m\left(A^C k + B^C\right)}{2}$$

*for $k$ such that $x_i^{**}(k) \in (0, a)$, where $A^C$, $B^C < 0$ are constants given in (10.21) and (10.22), respectively.*

Since $A^C < 0$, clearly $x_i^{**}(k)$ is increasing in $k$: the higher the stock of public bad, the higher the contribution to the abatement. This is the same as under non-cooperation. It can be verified that $x_i^{**}(k)$ is steeper than $x_i^{*}(k)$. As the stock of pollution increases, the response of contributors in the coordination scenario is stronger than that of their counterparts in the non-coordination scenario.

The equilibrium trajectory of the stock of pollution under coordination, $k^{**}(t)$, is the solution to the following first-order differential equation:

$$\frac{dk(t)}{dt} = an - mx_i^{**}(k(t)) - \delta k(t),$$

with $x_i^{**}(k)$ given in Proposition 10.4. It can be checked that

$$k^{**}(t) = k_{ss}^{**} + e^{-\sigma t}(k_0 - k_{ss}^{**}),$$

where

$$\sigma = \frac{\Gamma(m) - r}{2} > 0 \tag{10.23}$$

is the speed of convergence to $k_{ss}^{**}$, with

$$k_{ss}^{**} = \frac{m\left(1 + B^C\right) + 2a\left(n - m\right)}{2\delta - A^C m}. \tag{10.24}$$

It is immediate to verify that $\lim_{t\to\infty} k^{**}(t) = k_{ss}^{**}$. The following corollary can then be established.

**Corollary 10.2** *The vector of strategies ($x_1^{**}(k), x_2^{**}(k), ..., x_m^{**}(k)$) induces a trajectory of $k$ given by*

$$k^{**}(t) = k_{ss}^{**} + e^{-\sigma t}(k_0 - k_{ss}^{**})$$

*where $\sigma$ is the speed of convergence given in (10.23) and $k_{ss}^{**}$ is the (stable) steady state of $k$ under cooperative behavior given in (10.24).*

## 10.4 The Effects of an Increase in the Number of Abaters

In this section, we evaluate the impact of an increase in $m$ (keeping $n$ fixed) on total abatement, pollution, and social welfare in the static *versus* the dynamic game.

### *10.4.1 Static Effects*

#### 10.4.1.1 Non-coordination Scenario

From $x_i^*$ given in Proposition 10.1, the Nash equilibrium level of total contributions is equal to

$$X^*(m) = \frac{2b(k_0 + na) + 2a - 1}{2\left(b + \frac{1}{m}\right)}.$$

In the static non-cooperative scenario, total abatement is clearly increasing in $m$. Hence, the ex-post stock of pollution

$$k^*(m) = \frac{m + 2[k_0 + a(n-m)]}{2(bm+1)}$$

is decreasing in $m$. As a result of an increase in $m$, each contributor abates less, but the sum of abatement levels in the enlarged set of contributors turns out to be higher, which implies a lower pollution. The above discussion leads to the following proposition.

**Proposition 10.5** *$k^*(m') < k^*(m)$ with $m' > m$.*

As to social welfare, for analytical tractability, we set $n = 2$, $k_0 = 0$, and $b = 1$. From (10.4), we obtain the following expression:

$$W^*(m) = \frac{m + 4a(m-2)[a(5+2m) - 1]}{4(1+m)^2}.$$

We have

$$W^*(2) = \frac{1}{18},$$

and

$$W^*(1) = \frac{1 + 4a(1-7a)}{16}.$$

It follows that

$$W^*(2) - W^*(1) = \frac{(6a-1)(1+42a)}{144}.$$

The above expression is positive for $a > \widetilde{a} = 1/6$, which is required for $x_i^* > 0$ (see Proposition 10.1). We can then state the following proposition.

**Proposition 10.6** $W^*(2) > W^*(1)$.

In the static non-coordination scenario, an increase in the number of contributors from 1 to 2 is then welfare enhancing.

#### 10.4.1.2 Coordination Scenario

From $x_i^{**}$ given in Proposition 10.2, the level of total contributions under coordination is equal to

$$X^{**}(m) = \frac{2m\left[bk_0 m + a\left(1 + bmn\right)\right] - m}{2\left(1 + bm^2\right)}.$$

It can be easily verified that in the static coordination scenario, total abatement is increasing in $m$. Hence, the ex-post stock of pollution

$$k^{**}(m) = \frac{m + 2\left[k_0 + a\left(n - m\right)\right]}{2\left(bm^2 + 1\right)}$$

is decreasing in $m$. As a result of an increase in $m$, each contributor abates less, but the sum of abatement levels in the enlarged set of contributors turns out to be higher, which implies a lower pollution. This is the same as under the non-cooperative scenario. We can then state the following proposition.

**Proposition 10.7** $k^{**}\left(m'\right) < k^{**}(m)$ *with* $m' > m$.

As to social welfare, for analytical tractability, as done in the analysis of the non-coordination scenario, we set $n = 2$, $k_0 = 0$, and $b = 1$. From (10.4), we obtain the following expression:

$$W^{**}(m) = \frac{1}{4\left(1+m^2\right)^2}\Big\{m + 2m^2(m-1) + 4a\left\{2 + m\left[-5 - 2m(m-3)\right]\right\} + 4a^2(m-2)\left[5 + 2m\left(m^2 + m - 1\right)\right]\Big\}.$$

We have

$$W^{**}(2) = \frac{1}{10},$$

and

$$W^{**}(1) = W^{*}(1) = \frac{1 + 4a(1 - 7a)}{16}.$$

It follows that

$$W^{**}(2) - W^{**}(1) = \frac{3 - 20a + 140a^2}{80}.$$

The above expression is clearly positive for all $a$. We can then state the following proposition.

**Proposition 10.8** $W^{**}(2) > W^{**}(1)$.

In the static coordination scenario, similarly to what we observed in the non-coordination scenario, an increase in the number of contributors from 1 to 2 is then welfare enhancing.

## *10.4.2 Dynamic Effects*

### 10.4.2.1 Non-coordination Scenario

As in the static welfare analysis, for analytical tractability, we are going to consider the two cases $m = 1$ and $m = n = 2$, with $b = 1$. Let $k^*_{ss}(m)$ denote the steady-state stock of pollution with $m$ contributors. From (10.16), the steady-state stocks of pollution with $m = 1$ and $m = n = 2$ are given by

$$k^*_{ss}(1) = \frac{(1 + 2a)(r + \delta)}{2[1 + \delta(r + \delta)]}$$

and

$$k^*_{ss}(2) = \frac{5r + 4\delta + \sqrt{\Delta(2)}}{12 + \delta\left(5r + 4\delta + \sqrt{\Delta(2)}\right)},$$

respectively. It can be easily verified that $k^*_{ss}(2) - k^*_{ss}(1)$ is decreasing in $a$. Call $\widehat{a}$ the value of $a$ such that $k^*_{ss}(2) = k^*_{ss}(1)$. Hence, $k^*_{ss}(2) > k^*_{ss}(1)$ for $a < \widehat{a}$. The steady-state levels of contributions are given by

$$x^*_{ss}(1) = \frac{2a[2 + \delta(r + \delta)] - \delta(r + \delta)}{2[1 + \delta(r + \delta)]},$$

and

$$x_{ss}^{*}(2)=\frac{2a\left[12+\delta\left(5r+4\delta+\sqrt{\Delta(2)}\right)\right]-\delta\left(5r+4\delta+\sqrt{\Delta(2)}\right)}{24+2\delta\left(5r+4\delta+\sqrt{\Delta(2)}\right)}.$$

Observe that both $x_{ss}^{*}(1)$ and $x_{ss}^{*}(2)$ are increasing in $a$. Call $a_1$ and $a_2$ the values of $a$ such that $x_{ss}^{*}(1)=0$ and $x_{ss}^{*}(2)=0$, respectively. Clearly, we need $a>\max\{a_1,a_2\}$ for $x_{ss}^{*}(1), x_{ss}^{*}(2)>0$. It can be checked that $\widehat{a}>a_2>a_1$, with $\widehat{a}<1/2$ for $\delta>1/\sqrt{5}$. Hence, $\widehat{a}$ is admissible. A sufficiently large $\delta$ also guarantees that $x_{ss}^{*}(1), x_{ss}^{*}(2)<a$. The following proposition can then be established.

**Proposition 10.9** *There exists $\widehat{a}$ such that $k_{ss}^{*}(2)\geq$ (resp. $<$) $k_{ss}^{*}(1)$ for $a\leq$ (resp. $>$) $\widehat{a}$.*

If $a$ is below a certain threshold, $\widehat{a}$, an increase in the number of contributors from 1 to 2 leads to an increase in the steady-state stock of pollution. The intuition is as follows. Observe that the smaller $a$, the greater the marginal loss of direct utility caused by a given abatement level $x_i$ (because, for any given $x_i$, the marginal utility of consuming $a-x_i$ is higher when $a$ is lower). Therefore, when $a$ is small, a contributor's best reply to an increase in the sum of contributions by others tends to be a big reduction in her/his own contribution. This can lead to the result that an increase in the number of contributors induces a decrease in total abatement and hence an increase in $k_{ss}^{*}$. One may ask why there is no static counterpart to the result established in Proposition 10.5. The key for the answer is that in a dynamic model, each agent expects that if she/he increases her/his emission today, the pollution stock will be bigger tomorrow, which would in turn induce other agents to emit somewhat less tomorrow than otherwise; this dynamic strategic consideration may give her/him an incentive to undertake less abatement today when she/he learns that the number of contributors has increased. In a static model, by definition, such dynamic strategic considerations do not exist.

From (10.8), the steady-state level of social welfare is given by

$$w_{ss}^{*}(m)=a\left(n+2mx_{ss}^{*}\right)-na^{2}-bn\left(k_{ss}^{*}\right)^{2}-mx_{ss}^{*}\left(1+x_{ss}^{*}\right).$$

In the two cases under consideration, i.e., $m=1$ and $m=n=2$, letting $b=1$ we have

$$w_{ss}^{*}(1)=\frac{(r+\delta)\left[\delta^{3}+r\left(\delta^{2}-2\right)\right](1+4a)-4a^{2}\left\{2+(r+\delta)\left[4\delta+\delta^{3}+r\left(2+\delta^{2}\right)\right]\right\}}{4\left[1+\delta(r+\delta)\right]^{2}},$$

and

$$w_{ss}^{*}(2)=\frac{13r^{2}\left(\delta^{2}-4\right)-20r\sqrt{\Delta}+2\left(\delta^{2}-1\right)\left(12+5\delta^{2}+2\delta\sqrt{\Delta}\right)+r\delta\left(22\delta^{2}-28+5\delta\sqrt{\Delta}\right)}{\left[12+\delta\left(5r+4\delta+\sqrt{\Delta}\right)\right]^{2}}.$$

Clearly, $w_{ss}^{*}(2) - w_{ss}^{*}(1)$ is convex in $a$. Call $\widetilde{a}_1$ and $\widetilde{a}_2$ with $\widetilde{a}_1 > \widetilde{a}_2$ the two roots of $w_{ss}^{*}(2) - w_{ss}^{*}(1)$. It can be checked that $\widetilde{a}_1 \in (0, 1/2)$, whereas $\widetilde{a}_2$ can be either positive or negative (or nil). Hence, one can always find an $a \in (\max\{0, \widetilde{a}_2\}, \widetilde{a}_1)$ such that $w_{ss}^{*}(2) < w_{ss}^{*}(1)$. Take for instance $\delta = 2$ and $r = 0.1$. For these parameter values, we have $\widetilde{a}_1 = 0.362887$, $\widetilde{a}_2 = 0.104286$, and $\widehat{a} = 0.364189$. With $a = 0.35 \in (\max\{0, \widetilde{a}_2\}, \widetilde{a}_1)$ we expect $w_{ss}^{*}(2) < w_{ss}^{*}(1)$. Indeed,

$$w_{ss}^{*}(2) = 0.210796 < w_{ss}^{*}(1) = 0.215113.$$

Moreover, since $a < \widehat{a}$,

$$k_{ss}^{*}(2) = 0.349 > k_{ss}^{*}(1) = 0.343269,$$

and

$$2x_{ss}^{*}(2) = 0.00200083 < x_{ss}^{*}(1) = 0.0134615.$$

At the steady state, total abatement is higher and the stock of pollution is lower with one abater rather than two abaters. Interestingly, in the non-coordination scenario, in contrast with the static analysis, an increase in the number of abaters can worsen social welfare. We now take $a = 0.37 \notin (\max\{0, \widetilde{a}_2\}, \widetilde{a}_1)$. We have

$$w_{ss}^{*}(2) = 0.210796 > w_{ss}^{*}(1) = 0.20822.$$

Moreover, since $a > \widehat{a}$,

$$k_{ss}^{*}(2) = 0.349 < k_{ss}^{*}(1) = 0.351346,$$

and

$$2x_{ss}^{*}(2) = 0.0420008 > x_{ss}^{*}(1) = 0.0373077.$$

The above discussion leads to the following proposition.

**Proposition 10.10** *There exist* $\widetilde{a}_1$ *and* $\widetilde{a}_2$ *such that* $w_{ss}^{*}(2) \leq$ *(resp.* $>$*)* $w_{ss}^{*}(1)$ *for* $a \in$ *(resp.* $\notin$*)* $[\max\{0, \widetilde{a}_2\}, \widetilde{a}_1]$.

#### 10.4.2.2 Coordination Scenario

From (10.16), the steady-state stocks of pollution with $m = 1$ and $m = n = 2$ are given by

$$k_{ss}^{**}(1) = k_{ss}^{*}(1) = \frac{(1 + 2a)(r + \delta)}{2[1 + \delta(r + \delta)]}$$

and

$$k_{ss}^{**}(2) = \frac{8 - r^2 + r\delta + 2\delta^2 + (r+\delta)\sqrt{\Gamma(2)}}{\left(4 + r\delta + \delta^2\right)\left(2\delta - r + \sqrt{\Gamma(2)}\right)},$$

respectively. It can be easily verified that $k_{ss}^{*}(2) - k_{ss}^{*}(1)$ is decreasing in $a$. Call $\widehat{a}^C$ the value of $a$ such that $k_{ss}^{**}(2) = k_{ss}^{**}(1)$. Hence, $k_{ss}^{**}(2) > k_{ss}^{**}(1)$ for $a < \widehat{a}^C$. The steady-state levels of contributions are given by

$$x_{ss}^{**}(1) = x_{ss}^{*}(1) = \frac{2a\left[2 + \delta(r+\delta)\right] - \delta(r+\delta)}{2\left[1 + \delta(r+\delta)\right]},$$

and

$$x_{ss}^{**}(2) = \frac{8 + 2a(2 + r\delta)\left[4 + \delta(r+\delta)\right] - \delta\left\{\delta\left[r(r+\delta) - 2\right] + 2\sqrt{\Gamma(2)}\right\}}{2(2 + r\delta)\left[4 + \delta(r+\delta)\right]}.$$

Observe that both $x_{ss}^{**}(1)$ and $x_{ss}^{**}(2)$ are increasing in $a$. Call $a_1^C$ and $a_2^C$ the values of $a$ such that $x_{ss}^{**}(1) = 0$ and $x_{ss}^{**}(2) = 0$, respectively. Clearly, we need $a > \max\{a_1^C, a_2^C\}$ for $x_{ss}^{**}(1), x_{ss}^{**}(2) > 0$. It can be checked that $\widehat{a}^C > a_1^C > a_2^C$, with $\widehat{a}^C < 1/2$ for $\delta > 1/\sqrt{2}$. Hence, $\widehat{a}^C$ is admissible. A sufficiently large $\delta$ also guarantees that $x_{ss}^{**}(1), x_{ss}^{**}(2) < a$. The following proposition can then be established.

**Proposition 10.11** *There exists* $\widehat{a}^C$ *such that* $k_{ss}^{**}(2) \geq$ *(resp.* $<$*)* $k_{ss}^{**}(1)$ *for* $a \leq$ *(resp.* $>$*)* $\widehat{a}^C$.

As in the non-coordination scenario, conditions exist under which an increase in the number of contributors from 1 to 2 leads to an increase in the steady-state stock of pollution.

From (10.8), the steady-state levels of social welfare with $m = 1$, $m = n = 2$ and $b = 1$ are given by

$$w_{ss}^{**}(1) = \frac{(r+\delta)\left[\delta^3 + r\left(\delta^2 - 2\right)\right](1+4a) - 4a^2\left\{2 + (r+\delta)\left[4\delta + \delta^3 + r\left(2 + \delta^2\right)\right]\right\}}{4\left[1 + \delta(r+\delta)\right]^2},$$

and

$$w_{ss}^{**}(2) = \frac{1}{2(2 + r\delta)^2\left[4 + \delta(r+\delta)\right]^2}\left\{r^4\delta^2\left(\delta^2 - 4\right)\right.$$
$$+4\left(-16 + 4\delta\sqrt{\Gamma(2)} - 7\delta^2\right)\left(4 + \delta^2\right)$$
$$+2r^3\delta\left(-12 + 2\delta^2 + \delta^4\right) + r^2\left(-40 - 8\delta\sqrt{\Gamma(2)} - 12\delta^2 + 12\delta^4 + \delta^6\right)$$
$$\left.+\,4r\left[\sqrt{\Gamma(2)}\left(4\delta^2 - 6\right) + \delta\left(-28 - 6\delta^2 + \delta^4\right)\right]\right\}.$$

Clearly, $w_{ss}^{**}(2) - w_{ss}^{**}(1)$ is convex in $a$. Call $\widetilde{a}_1^C$ and $\widetilde{a}_2^C$ with $\widetilde{a}_1^C > \widetilde{a}_2^C$ the two roots of $w_{ss}^{**}(2) - w_{ss}^{**}(1)$. It can be checked that $\widetilde{a}_1^C > 0$ and $\widetilde{a}_2^C < 1/2$. For the parameter values previously considered, i.e., $\delta = 2$ and $r = 0.1$, we have $\widetilde{a}_1^C = 0.567192$, $\widetilde{a}_2^C = -0.100019$, and $\widehat{a}^C = 0.38506$. With $a = 0.35 \in (\max\{0, \widetilde{a}_2^C\}, \min\{\widetilde{a}_1^C, 1/2\})$ we then expect $w_{ss}^{**}(2) < w_{ss}^{**}(1)$. Indeed,

$$w_{ss}^{**}(2) = 0.0818763 < w_{ss}^{**}(1) = 0.215113.$$

Moreover, since $a < \widehat{a}^C$,

$$k_{ss}^{**}(2) = 0.357428 > k_{ss}^{**}(1) = 0.343269,$$

and

$$2x_{ss}^{**}(2) = 0.270288 > x_{ss}^{**}(1) = 0.0134615.$$

At the steady state, both total abatement and the stock of pollution are lower with one abater rather than two abaters. Interestingly, in the coordination scenario, in contrast with the static analysis, an increase in the number of abaters can worsen social welfare. The above discussion leads to the following proposition.

**Proposition 10.12** *There exist $\widetilde{a}_1^C$ and $\widetilde{a}_2^C$ such that $w_{ss}^{*}(2) \leq$ (resp. $>$) $w_{ss}^{*}(1)$ for $a \in$ (resp. $\notin$)* $[\max\{0, \widetilde{a}_2^C\}, \min\{\widetilde{a}_1^C, 1/2\}]$.

## 10.5 Concluding Remarks

We have analyzed both a static and a dynamic game of voluntary abatement of a public bad. In each game, an exogenously given coalition of abaters coexists with a group of free riders. Countries/agents in the coalition agree on the need for taking action to reduce the public bad. However, the level of individual contributions depends on the mode of cooperation. Under tight cooperation, each abating country/agent contributes so as to maximize the sum of payoffs of all the participating countries/agents in the coalition. Under loose cooperation, instead, the individual level of contributions is determined in a decentralized fashion with contributors considering only their own payoffs and taking as given the contributions of the other coalition members. One of the striking results of our analysis is that an increase in the size of the coalition can increase the stock of pollution and worsen social welfare in the dynamic game, whereas, in the static game, more contributors are always associated with lower pollution and higher social welfare. A policy implication of this finding is that conditions exist under which smaller coalitions turn out to be more efficient than bigger ones. Specifically, we have shown that an increase in the size of the coalition from 1 to 2 leads to an increase in the steady-state stock of pollution and to a decrease in social welfare provided that the "business-as-usual" level of output belongs to a certain range. In the context of IEAs, our dynamic analysis suggests that evaluating the success or the failure of an agreement based on the number of participating countries alone can lead to erroneous conclusions.

Possible extensions of our framework include (i) relaxing the assumption of full cooperation by considering a coefficient of cooperation ranging from zero to one (see, for instance, Vives, 2008; Colombo & Labrecciosa, 2018; Colombo et al., 2022), (ii) allowing for agents' heterogeneity (see McGinty, 2007; Pavlova & de Zeeuw, 2013), (iii) allowing for asymmetric information about agents' characteristics (see Bagwell & Staiger, 2005; Amador & Bagwell, 2013), and (iv) accounting for uncertainty and ambiguity about the evolution of the stock of public bad (see, for instance, Lemoine & Traeger, 2014, 2016).

## References


Amador, M., & Bagwell, K. (2013). The theory of optimal delegation with an application to tariff caps. *Econometrica, 81*, 1541–1599.

Andreoni, J. (1988). Privately provided public goods in a large economy: The limits of altruism. *Journal of Public Economics, 35*, 57–73.

d'Aspremont, C., Jacquemin, J., Gabszewicz, J., & Weymark, J. (1983). On the stability of collusive price leadership. *Canadian Journal of Economics, 16*, 17–25.

Atatüre, S. (2008). The historical roots of European Union: integration, characteristics, and responsibilities for the 21st century. *European Journal of Social Sciences. Eurojournal, 7*(2), 20.

Bagwell, K., & Staiger, R. W. (2005). Enforcement, private political pressure and the general agreement on tariffs and trade. *Journal of Legal Studies, 3*, 471–513.

Barrett, S. (1994). Self-enforcing international environmental agreements. *Oxford Economics Papers, 46*, 878–894.

Battaglini, M., Nunnari, S., & Palfrey, T. R. (2014). Dynamic free riding with irreversible investments. *American Economic Review, 104*, 2858–2871.

Benchekroun, H., & Long, N. V. (2008). The build-up of cooperative behavior among non-cooperative selfish agents. *Journal of Economic Behavior and Organization, 67*, 239–252.

Bergstrom, T. C., Blume, L. E., & Varian, H. R. (1986). On the private provision of public goods. *Journal of Public Economics, 29*, 25–49.

Bowen, R., Georgiadis, G., & Lambert, N. S. (2019). Collective choice in dynamic public good provision. *American Economic Journal: Microeconomics, 11*, 243–298.

Carraro, C., & Siniscalco, D., 1993. Strategies for the international protection of the environment. *Journal of Public Economics, 52*, 309–28.

Chamberlin, J. (1974). Provision of collective goods as a function of group size. *American Political Science Review, 65*, 707–716.

Colombo, L., & Labrecciosa, P. (2018). Consumer surplus-enhancing cooperation in a natural resource oligopoly. *Journal of Environmental Economics and Management, 92*, 185–193.

Colombo, L., Labrecciosa, P., & Long, N. V. (2019). A Dynamic analysis of climate change mitigation with endogenous number of contributors: Loose vs tight cooperation. Discussion Paper Series, Hitotsubashi Institute for Advanced Study, Hitotsubashi University, HIAS-E-92.

Colombo, L., Labrecciosa, P., & Long, N. V. (2022). A dynamic analysis of international environmental agreements under partial cooperation. *European Economic Review, 143*, 104036.

Cornes, R., & Sandler, T. (1986). *The theory of externalities, public goods and club goods* (2nd ed.). Cambridge University Press.

de Zeeuw, A. (2008). Dynamic effects on the stability of international environmental agreements. *Journal of Environmental Economics and Management, 55*, 163–174.
Eichner, T., & Pethig, R. (2013). Self-enforcing environmental agreements and international trade. *Journal of Public Economics, 102*, 37–50.
Fershtman, C., & Nitzan, S. (1991). Dynamic voluntary provision of public goods. *European Economic Review, 35*, 1057–1067.
Fujiwara, K., & Matsueda, N. (2009). Dynamic voluntary provision of public goods: An extension. *Journal of Public Economic Theory, 11*, 27–36.
Georgiadis, G. (2015). Project and Team Dynamics. *Review of Economic Studies, 82*, 187–218.
Georgiadis, G. (2017). Deadlines and infrequent monitoring in the dynamic provision of public goods. *Journal of Public Economics, 152*, 1–12.
Itaya, J., & Shimomura, K. (2001). A dynamic conjectural variations model in the private provision of public goods: a differential game approach. *Journal of Public Economics, 81*, 153–172.
Lemoine, D., & Traeger, C. (2014). Watch your step: Optimal policy in a tipping climate. *American Economic Journal: Economic Policy, 6*, 137–166.
Lemoine, D., & Traeger, C. (2016). Ambiguous tipping points. *Journal of Economic Behavior and Organization, 132*, 5–18.
Long, N. V., & Shimomura, K. (2007). Voluntary contributions to a public good: Non-neutrality results. *Pacific Economic Review, 12*, 153–170.
McGinty, M. (2007). International environmental agreements among asymmetric nations. *Oxford Economic Papers, 59*, 45–62.
Marx, L. M., & Matthews, S. A. (2000). Dynamic voluntary contribution to a public project. *Review of Economic Studies, 67*, 327–358.
Pavlova, Y., & de Zeeuw, A. (2013). Asymmetries in international environmental agreements. *Environmental and Development Economics, 18*, 51–68.
Rubio, S., & Ulph, A. (2006). Self-enforcing international environmental agreements revisited. *Oxford Economic Papers, 58*, 233–263.
Rubio, S. J., & Ulph, A. (2007). An infinite-horizon model of dynamic membership of international environmental agreements. *Journal of Environmental Economics and Management, 54*, 296–310.
Varian, H. R. (1994). Sequential contributions to public goods. *Journal of Public Economics, 53*, 165–186.
Vives, X. (2008). Innovation and competitive pressure. *Journal of Industrial Economics, 56*, 419–496.
Wirl, F. (1996). Dynamic voluntary provision of public goods: Extension to nonlinear strategies. *European Journal of Political Economy, 12*, 555–560.
Yanase, A. (2006). Dynamic voluntary provision of public goods and optimal steady-state subsidies. *Journal of Public Economic Theory, 8*, 171–179.

# Chapter 11
# Agroecology and Biodiversity: A Benchmark Dynamic Model

**Emmanuelle Augeraud-Véron, Raouf Boucekkine, and Rodolphe Desbordes**

**Abstract** Conventional agriculture not only neglects but also harms the ecosystem services provided by biodiversity, inducing a negative feedback loop. In a theoretical inspired by agroforestry ("agriculture with trees"), a common agroecological practice in developing countries, we highlight how the choice between expanding agricultural land and retaining forest land is shaped by the bidirectional relationship between agriculture and biodiversity as well as the utility derived from biodiversity consumption. The static case shows that a high stock of biodiversity may be deliberately maintained as long as the agroecological productivity effect is important enough. This result holds in the dynamic case. However, in the latter case, a large intertemporal discount rate can lead to total biodiversity loss along with the full collapse of the economy. Another key implication of our model, among other results, is that the effect of a shift of consumer preferences toward agricultural goods (instead of biodiversity goods) on the biodiversity stock is much more ambiguous in the dynamic case than in the static case, depending on the strength on the agroecological productivity effect. These results have profound implications for biodiversity conservation.

**Keywords** Agriculture · Agroecology · Agroforestry · Biodiversity · Dynamic · Feedback loop · Static · Sustainability

---

E. Augeraud-Véron
Bordeaux School of Economics, Bordeaux, France

R. Boucekkine (✉)
Aix-Marseille School of Economics, Aix-Marseille University and CNRS, Marseille, France
e-mail: raouf.boucekkine@univ-amu.fr

R. Desbordes
SKEMA Business School – Université Côte d'Azur, Paris, France

 
F. El Ouardighi, G. Feichtinger (eds.), *The Unaffordable Price of Static Decision-making Models*, International Series in Operations Research & Management Science 365, https://doi.org/10.1007/978-3-031-88638-6_11

## 11.1 Introduction

Over the period 1960–2019, land use change has affected almost a third of global land area and has involved substantial expansion of agricultural land (Winkler et al., 2021; Potapov et al., 2022a,b). These land transformations have considerably harmed biodiversity. Newbold et al. (2015) estimate that, in 2005, species richness and abundance have globally declined by 11–14%, with much stronger local impacts (40–50%) for intensively exploited agricultural land characterized by monocultures, heavy use of synthetic fertilizers and pesticides, and mechanization. This biodiversity loss implies both the neglect and degradation of key benefits provided by ecosystems to agriculture (Zhang et al., 2007), notably in terms of supporting services (e.g., soil structure and fertility) and regulating services (e.g., pollination and pest control). Negative feedback loops may thus emerge between conventional agricultural practices and ecosystem services, progressively harming agricultural output, especially if any negative effects cannot be compensated by external inputs (Foley et al., 2005; Ortiz et al., 2021).

In response to this trade-off between short-run gains and long-run losses as well as the large environmental and human costs (e.g., water pollution, freshwater scarcity, emerging infectious diseases, climate change) associated with food production, more sustainable farming approaches have been proposed (Tscharntke et al., 2012; Bommarco et al., 2013; Nair, 2014; Tittonell et al., 2016; Garibaldi et al., 2017). This "biodiversity"-based agriculture aims at harnessing, rather than substituting, the ecosystem services provided by biodiversity. It aims at optimally integrating the biological and ecological processes within the agroecosystem while minimizing the use of external non-renewable inputs that cause environmental or human harm (Pretty, 2008; Kremen et al., 2012; Wezel et al., 2014; Duru et al., 2015). Among the various ESR (efficiency/substitution/redesign) agroecological practices, related to crop or landscape management, which may be adopted (Wezel et al., 2014), agroforestry ("agriculture with trees") is relatively widespread in tropical and developing countries. Defined as "the purposeful growing or deliberate retention of trees with crops and/or animals in interacting combinations for multiple products or benefits from the same management unit" (Nair et al., 2021), agroforestry is estimated to involve 43% of all agricultural land (at least 10% tree cover) and may provide direct subsistence to about a billion people (Zomer et al., 2014, 2016). Several meta-analyses have highlighted that agroforestry, in comparison to a less diverse agrosystem and even to other crop diversification strategies, tends to increase associated biodiversity, raise soil quality and fertility, improve pest and diseases control as well as pollinator abundance, and generate yield improvements (Pumariño et al., 2015; Torralba et al., 2016; Dainese et al., 2019; Kuyah et al., 2019; Staton et al., 2019; Udawatta et al., 2019; Beillouin et al., 2021; Baier et al., 2023; Centeno-Alvarado et al., 2023).[1] These benefits have induced various authors

[1] The biodiversity improvements do not necessarily mean that the environment is not affected by agroforestry. For example, Chaudhary et al. (2016) find that agroforestry, while the least

and international organizations, such as the IPCC (The Intergovernmental Panel on Climate Change), to advocate greater use of agroforestry to meet a large number of key sustainable development goals (Waldron et al., 2017; van Noordwijk et al., 2018; Shukla et al., 2019).

Our paper contributes to the field of agroforestry, and more broadly agroecology, by theoretically highlighting the relatively ignored negative "feedback loop" existing between agricultural expansion and ecosystem services. Indeed, in their review, (Ortiz et al., 2021) consider that the bidirectional relationship between agricultural production and biodiversity is not sufficiently considered and understood. We build up a benchmark non-spatial dynamic model incorporating this bidirectional relationship allowing to assess in depth the potential long-term implications of a well-defined feedback effect of biodiversity on the productivity of the agricultural sector in accordance with the empirical literature surveyed above. We call this effect the *agroecological productivity effect*. Other potentially important ingredients are also incorporated.

The model runs as follows. An agricultural economy produces an agricultural good with a simple Cobb-Douglas technology with land as a principal input (labor is normalized to 1). No storage of this good is possible, so it's consumed entirely at every period of time. The key addition to this otherwise standard agricultural economy model is a feedback effect of the biodiversity stock on the productivity of the agricultural sector: the larger this stock, the more productive is the agricultural land (agroecological productivity effect or boost). Biodiversity dynamics are driven by a simple accumulation law of motion depending on the initial stock of biodiversity, the remaining forest (non-agricultural) land, and harvesting of this biodiversity by the population. We identify harvesting with bush (or wild) food consumption. The representative farmer maximizes an intertemporal discounted flow of utility derived from consuming both the agricultural and wild good subject to biodiversity dynamics. It turns out by construction that the farmer has two independent controls, the land use variable (precisely, the size of the agricultural vs. forest areas) and the consumption of bush food.

In order to have a first idea on the qualitative implications of the agroecological productivity effect, we study a preliminary static optimization counterpart. We highlight the main effects at work in the general case (that is under general functional specifications for the feedback effect and the instantaneous utility). However, to derive analytical results, we specialize in Cobb-Douglas utility functions and linear feedback effects. We solve the corresponding optimization problem and highlight the main qualitative properties of the optimal controls, notably through comparative statics with respect to three key parameters: the regeneration rate of biodiversity, the elasticity of instantaneous utility with respect to the agricultural good (which allows to capture to which extent the farmer "prefers" the agricultural good with respect to the wild one), and a direct indicator of the agroecological productivity

detrimental non-timber-producing management regimes, still reduces on average species richness found in nearby natural forests by 32%.

effect. We essentially enhance the role of the feedback effects in this static case. We then move to the optimal control problem and perform the typical mathematical treatment, establishing the sufficiency of the first-order Pontryagin conditions, the existence and uniqueness of the steady state, and its local (saddle point) stability. We end our analysis with a careful analysis of the comparative statics resulting at the optimal steady state with respect to the three key parameters listed above.

Besides the methodological value-added deriving from the comparison between the static and dynamic frames, several highly interesting economic results have been identified. We mention two in this Introduction. First of all, we highlight the importance of the time discounting rate in the presence of the agroecological productivity effect. Remarkably enough, with our canonical specifications, we show that the stock of biodiversity converges to 0 when the discount rate becomes infinitely large. Due to the agroecological feedback effect, the whole economy goes to extinction as the vanishing stock of biodiversity also leads to the collapse of the agricultural sector in this case. Second, we establish several properties at the steady state of the dynamic model, which are quite interesting from the policy point of view. For example, we prove that as the agricultural society preferences move away from bush food consumption (may be following government campaigning), the stock of biodiversity is raised only if the agroecological productivity effect is strong enough, further leading to a virtuous circle. This implies that campaigning with a biodiversity conservation policy is doomed to failure.

The paper is organized as follows. Section 11.2 presents the model and the associated optimal control problem. Section 11.3 constructs a static counterpart and studies its qualitative properties at the optimum. Section 11.4 solves the optimal control problem and derives its qualitative properties at the steady state, with comparison to the static case counterpart, enhancing the interaction between the feedback effects inherent in the model and the intertemporal mechanisms. Section 11.5 concludes.

## 11.2 The Model

We consider a one-sector agricultural economy where the agricultural good is produced according to the following general technology:

$$Y = F\left(A, T(B)L\right),$$

where $A$ is the size of the agricultural land, $L$ is the available labor force, and $T(B)$ is labor-saving technology which depends on the amount of biodiversity available. This specification fits several contexts: in the case of agroecology, nature services increase the productivity of agriculture; in such a case, $T(.)$ is a non-decreasing function of biodiversity.

Moreover, we normalize initial labor resources and forest land surface to 1. We do not model space explicitly. Denoting by $f$ the fraction of forest land preserved such that $A = 1 - f$, we assume a Cobb-Douglas agricultural production function:

$$Y = (T(B))^{1-\alpha}\ (1-f)^{\alpha}, \tag{11.1}$$

with $0 < \alpha < 1$. The agricultural production is entirely consumed at any date, that is:

$$Y = C_A, \tag{11.2}$$

where $C_A$ is consumption of the agricultural good or domestic food. The individual (or the economy) can also directly consume comestible biodiversity (animals or vegetables) or bush food, denoted $C_B$, which may or may not lead to further deforestation. Both cases may be considered. In this simple model, we assume that consumption of bush food does not involve any significant cost in terms of deforestation or in terms of labor.[2]

**Biodiversity Dynamics** We now come to the link between agricultural activity and the evolution of biodiversity over time. We set the following law of motion:

$$\dot{B} = R\,(f, B)\ B - C_B,$$

where $R(f, B)$ is the natural regeneration rate of biodiversity, which is here assumed to depend on the extent of forests and the existing stock of biodiversity. $B(0) > 0$ is given. In this paper, we shall choose the following specification:

$$\dot{B} = \beta\ B^{1-\theta} f^{\theta} - C_B, \tag{11.3}$$

with $0 < \theta < 1$. The latter equation can be rewritten in the more meaningful way:

$$\frac{\dot{B}}{B} = \beta\ \left(\frac{f}{B}\right)^{\theta} - \frac{C_B}{B}.$$

**The Optimal Control Problem** The decision maker chooses the trajectories of $\{f, C_A, C_B\}$ from $t = 0$ aimed at maximizing the following intertemporal utility function:

$$\int_0^{\infty} U\,(C_A, C_B)\ e^{-\delta\, t}\ dt \tag{11.4}$$

[2] One could legitimately argue that hunting, for example, is time-consuming and diverts labor from agriculture. Adding labor allocation between agriculture and consumption of biodiversity will not alter the main correlations between land use, biodiversity, and zoonoses we target in this paper.

subject to Eqs. (11.1), (11.2), and (11.3), with $B(0)$ given. $\delta > 0$ is the time discounting rate. We shall consider standard utility functions, namely, strictly increasing in each of the arguments, concave in $(C_A, C_B)$, and checking the Inada conditions for each argument of the function. Using Eqs. (11.1)–(11.2), one can substitute for $C_A$ in the objective function, leading to an optimal control problem with two controls $\{f, C_B\}$ and one state $B$. More precisely, the optimal control problem can be finally rewritten as:

$$\max_{f,c_B} \int_0^{\infty} U\left([T(B)]^{1-\alpha}\,(1-f)^{\alpha}, C_B\right)\, e^{-\delta t}\, dt$$

subject to

$$\dot{B} = \beta\, B^{1-\theta}\, f^{\theta} - C_B,$$

with $B(0) = B_0$ given and the usual positivity conditions (plus $f \leq 1$).

## 11.3 A Static Counterpart Model

### *11.3.1 The Counterpart Static Optimization Problem*

Consider the same problem within a single period, starting with a biodiversity level, $B_0$ normalized to 1 for simplicity. Then the biodiversity level within the period is given by (modulo constants inherent in the discretization step, irrelevant for the qualitative properties of the optimal solutions if any):

$$B = \beta\, f^{\theta} - C_B,$$

leading to the following one-period optimization problem:

$$\max_{f,c_B} U\left(\left[T\left\{\beta\, f^{\theta} - C_B\right\}\right]^{1-\alpha}\,(1-f)^{\alpha}, C_B\right). \tag{11.5}$$

The static optimization problem (11.5) is highly nontrivial for general functions $T(.)$ and $U(.,.)$. The first-order necessary optimality conditions write as follows (with $U_i(.,.)$ the partial derivative of $U$ with respect to its $i$th argument, $i = 1, 2$):

$$U_2(.,.) = (1-\alpha)U_1(.,.)T'(.)\,[T(.)]^{-\alpha}\,(1-f)^{\alpha},$$

and

$$(1-\alpha)\beta\theta U_1(.,.)T'^{-\alpha}(1-f)^{\alpha} f^{\theta-1} = \alpha U_1(.,.)T^{1-\alpha}(1-f)^{\alpha-1},$$

where $U_1$ and $U_2$ are evaluated at $\left(\left[T\left\{\beta f^{\theta} - C_B\right\}\right]^{1-\alpha} (1-f)^{\alpha}, C_B\right)$ and $T$ and $T'$ are evaluated at $\beta\ f^{\theta}\ - C_B$. The first equation is the necessary optimality condition with respect to $C_B$: it equalizes the marginal benefit from harvesting biodiversity ($U_2$) to the corresponding marginal cost (the right-hand side of the equation) reflecting the marginal loss in biodiversity impacting productivity in the agricultural sector, which in turn affects negatively consumption of the agricultural good. The second equation is the necessary optimality condition with respect to $f$: again it equalizes the marginal benefit from increasing the forest surface through increasing biodiversity (thus productivity and ultimately the consumption of the agricultural good) to the corresponding marginal cost, which is simply the one decreasing the latter consumption as the resulting agricultural land shrinks. Both conditions make perfect sense as they accurately reflect the economic trade-offs at work in our model. The two equations are quite intricate algebraically in the general case; even after simplification of the second equation, we shall refer to these final formulations below:

$$U_2(.,.) = (1-\alpha)U_1(.,.)T'(.)\,[T(.)]^{-\alpha}\,(1-f)^{\alpha} \tag{11.6}$$

and

$$\frac{T'(.)}{T(.)} = \frac{\alpha}{(1-\alpha)\beta\theta}\frac{f^{1-\theta}}{1-f}. \tag{11.7}$$

Equation (11.7) is interesting: it gives the growth rate of agricultural productivity as a highly nonlinear increasing function of forest land in the agroecological context we are studying.

Given the objectives of this paper, we shall concentrate on functional specifications which deliver closed-form solutions for the static problem and the dynamic extension (at the steady state). The analytical case specifications given below will therefore also serve for the latter in Sect. 11.4.

**An Analytically Tractable Case** A full analytical solution to the static problem is obtained with the following specifications:

1. $U(x, y) = x^{\gamma}\ y^{1-\gamma}$, with $0 < \gamma < 1$.
2. $T(x) = x$.

Clearly the two specifications are highly "stylized": the agricultural and non-agricultural goods may show in some practical cases a strong form of complementarity, and the relationship between productivity and the amount of biodiversity is probably nonlinear. In our model, the amount of biodiversity will remain bounded over optimal trajectories, so the postulated linearity of productivity is not so problematic. With these specifications, the problem (11.5) becomes

$$\max_{f, C_B} \left[\beta f^{\theta} - C_B\right]^{\gamma(1-\alpha)} (1-f)^{\gamma\alpha} C_B^{1-\gamma}. \tag{11.8}$$

Using the first-order optimality conditions (11.6)–(11.7) with the specifications above yields:

$$C_B = \frac{(1-\gamma)}{\gamma(1-\alpha)} \left(\beta f^{\theta} - C_B\right), \tag{11.9}$$

and

$$\frac{1}{\beta f^{\theta} - C_B} = \frac{\alpha}{(1-\alpha)\beta\theta} \frac{f^{1-\theta}}{1-f}. \tag{11.10}$$

The following proposition summarizes the main results.

**Proposition 11.1** *The maximization problem (11.8) admits a unique solution* $\left(f^*, C_B^*\right)$ *such that:*

- $f^* = \frac{(1-\alpha\gamma)\theta}{\gamma\alpha+\theta(1-\alpha\gamma)}$ *(so* $f^* < 1$*).*
- $C_B^* = \frac{(1-\gamma)\beta}{1-\alpha\gamma} (f^*)^{\theta}$.
- *Consequently, one gets:* $B^* = \beta \frac{\gamma(1-\alpha)}{1-\alpha\gamma} (f^*)^{\theta}$, *and* $C_A^* = (B^*)^{1-\alpha} (1-f^*)^{\alpha}$.

***Proof*** To prove this latter point, first-order derivatives are given as follows.

$$\begin{aligned}
\frac{\partial U}{\partial C_B} &= \left(-\gamma(1-\alpha)\ C_B + (1-\gamma)\left[\beta f^{\theta} - C_B\right]\right) \\
&\quad (1-f)^{\gamma\alpha} \left[\beta f^{\theta} - C_B\right]^{\gamma(1-\alpha)-1} C_B^{-\gamma} \\
\frac{\partial U}{\partial f} &= \left((1-\alpha)\beta\theta f^{\theta-1} (1-f) - \alpha \left[\beta f^{\theta} - C_B\right]\right) \gamma \left[\beta f^{\theta} - C_B\right]^{\gamma(1-\alpha)-1} \\
&\quad (1-f)^{\gamma\alpha-1} C_B^{1-\gamma}.
\end{aligned}$$

Consistently the optimum yields:

$$C_B = \frac{(1-\gamma)}{\gamma(1-\alpha)} \left[\beta f^{\theta} - C_B\right]$$

$$\frac{1}{\beta f^{\theta} - C_B} = \frac{\alpha f^{1-\theta}}{(1-\alpha)\beta\theta\,(1-f)}$$

$$C_B = \frac{1-\gamma}{1-\alpha\gamma} \beta f^{\theta}$$

$$\frac{(1-\alpha\gamma)\,\theta}{\gamma\alpha+\theta\,(1-\alpha\gamma)}=f.$$

Let us denote $\pi_c = (1-f^*)^{\gamma\alpha}\left[\beta f^{*\theta}-C_B^*\right]^{\gamma(1-\alpha)-1} C_B^{*-\gamma}, \pi_f = \gamma\left[\beta f^{*\theta}-C_B^*\right]^{\gamma(1-\alpha)-1}(1-f^*)^{\gamma\alpha-1}\, C_B^{*1-\gamma}$. The evaluation of the Hessian at steady state is given as follows:

$$\begin{bmatrix} (\alpha\gamma-1)\,\pi_c & -\beta\theta\left(\frac{-\theta(\alpha\gamma-1)}{\alpha\gamma-\theta(\alpha\gamma-1)}\right)^{(\theta-1)}(\gamma-1)\,\pi_c \\ \alpha\pi_f & -\beta\theta\left(\frac{-\theta(\alpha\gamma-1)}{\alpha\gamma-\theta(\alpha\gamma-1)}\right)^{\theta-2}\frac{\alpha^2\gamma\theta-\alpha^2\gamma-2\alpha\gamma\theta+\alpha\gamma+\theta}{\alpha\gamma-\theta(\alpha\gamma-1)}\pi_f \end{bmatrix}$$

$$\Delta=\Bigg[(\alpha\gamma-1)\left(-\beta\theta\left(\frac{-\theta\,(\alpha\gamma-1)}{\alpha\gamma-\theta\,(\alpha\gamma-1)}\right)^{\theta-2}\frac{\alpha^2\gamma\theta-\alpha^2\gamma-2\alpha\gamma\theta+\alpha\gamma+\theta}{\alpha\gamma-\theta\,(\alpha\gamma-1)}\right)$$

$$+\beta\theta\left(\frac{-\theta\,(\alpha\gamma-1)}{\alpha\gamma-\theta\,(\alpha\gamma-1)}\right)^{(\theta-1)}(\gamma-1)\,\alpha\Bigg]\left(\pi_c\pi_f\right)^2$$

$$=\Big[(\alpha\gamma-1)\left(-\left(\alpha^2\gamma\theta-\alpha^2\gamma-2\alpha\gamma\theta+\alpha\gamma+\theta\right)\right)$$

$$+(-\theta\,(\alpha\gamma-1))\,(\gamma-1)\,\alpha\Big]\left(\frac{-\theta\,(\alpha\gamma-1)}{\alpha\gamma-\theta\,(\alpha\gamma-1)}\right)^{\theta-2}\frac{\beta\theta}{\alpha\gamma-\theta\,(\alpha\gamma-1)}\left(\pi_c\pi_f\right)^2.$$

$\Delta$ has the same sign as $d$, given by:

$$\begin{aligned} d &= \Big[(\alpha\gamma-1)\left(-\left(\alpha^2\gamma\theta-\alpha^2\gamma-2\alpha\gamma\theta+\alpha\gamma+\theta\right)\right) \\ &\quad +(-\theta\,(\alpha\gamma-1))\,(\gamma-1)\,\alpha\Big] \\ &= (1-\alpha)\,(1-\alpha\gamma)\,(\theta\,(1-\alpha\gamma)+\alpha\gamma)\,. \end{aligned}$$

Thus, $\Delta > 0$. Moreover, $(\alpha\gamma-1) < 0$; thus, the optimum is a maximum. ■

We study the economic implications of Proposition 11.1 just below.

### 11.3.2 Economic Properties

To explore the economic properties of the optimal static solution, we will concentrate on three key parameters: the regeneration rate of biodiversity, $\beta$; the elasticity of instantaneous utility with respect to the agricultural good, $\gamma$; and a direct indicator of the agroecological productivity effect, $1-\alpha$. Indeed, $1-\alpha$ is the elasticity of the agricultural output with respect to biodiversity. The three parameters above "quantify" the three important ingredients of the paper: the biodiversity growth rate, the agroecological engine of agricultural production, and the arbitrage between the agricultural good and bush food (inherent in the preferences of the consumers).

The proposition below gives the comparative statics of the optimal solution for three key variables: the agricultural land, the biodiversity stock, and the amount of biodiversity harvested (to be consumed). The three variables show up in the biodiversity equation (static and dynamic), which is the central one for the purpose of our analysis as outlined above. We get the following economic picture.

**Proposition 11.2** *The following comparative statics hold:*

- $\frac{\partial f^*}{\partial \beta} = 0$, $\frac{\partial f^*}{\partial \alpha} < 0$, *and* $\frac{\partial f^*}{\partial \gamma} < 0$.
- $\frac{\partial C_B^*}{\partial \beta} > 0$, $\frac{\partial C_B^*}{\partial \gamma} < 0$, *and* $\frac{\partial C_B^*}{\partial \alpha} > 0$.
- *It follows that* $\frac{\partial B^*}{\partial \beta} > 0$, $\frac{\partial B^*}{\partial \alpha} < 0$ *and* $\frac{\partial B^*}{\partial \gamma} > 0$.

The computations deliver indeed the following results for the partial derivatives needed and the resulting signs. For forest land surface, one gets the following direct results:

$$\frac{\partial f^*}{\partial \alpha} = -\frac{\gamma\theta}{(\gamma\alpha + \theta(1-\alpha\gamma))^2} < 0,$$

$$\frac{\partial f^*}{\partial \gamma} = -\frac{\alpha\theta}{(\gamma\alpha + \theta(1-\alpha\gamma))^2} < 0.$$

Things are more involved for the stock of biodiversity, in particular for parameter $\gamma$:

$$\frac{\partial B^*}{B^*\partial \alpha} = -\frac{(1-\gamma)}{(1-\alpha)(1-\alpha\gamma)} - \frac{\gamma}{(1-\alpha\gamma)(\gamma\alpha + \theta(1-\alpha\gamma))} < 0,$$

$$\frac{\partial B^*}{B^*\partial \beta} = \frac{1}{\beta} > 0,$$

$$\begin{aligned}\frac{\partial B^*}{B^*\partial \gamma} &= \frac{1}{\gamma(1-\alpha\gamma)} - \frac{\alpha\theta}{(1-\alpha\gamma)(\gamma\alpha + \theta(1-\alpha\gamma))} \\ &= (\gamma\alpha + \theta(1-\alpha\gamma) - \gamma\alpha\theta)\frac{1}{\gamma(1-\alpha\gamma)(\gamma\alpha + \theta(1-\alpha\gamma))} \\ &= (\gamma\alpha + \theta(1-2\alpha\gamma))\frac{1}{\gamma(1-\alpha\gamma)(\gamma\alpha + \theta(1-\alpha\gamma))}.\end{aligned}$$

However, it can be seen that $\gamma\alpha + \theta(1-2\alpha\gamma) > 0$. Thus, $\frac{\partial B^*}{B^*\partial \gamma} > 0$.
The comparative statics of wild goods consumption is more straightforward:

$$\frac{\partial C^*}{C^*\partial\alpha} = \frac{\gamma^2\alpha(1-\theta)}{(1-\alpha\gamma)(\gamma\alpha+\theta(1-\alpha\gamma))} > 0$$
$$\frac{\partial C^*}{C^*\partial\beta} = \frac{1}{\beta} > 0,$$
$$\frac{\partial C^*}{C^*\partial\gamma} = \frac{1}{(1-\alpha\gamma)}\left(\frac{-(1-\alpha)}{(1-\gamma)} - \frac{\alpha}{(\gamma\alpha+\theta(1-\alpha\gamma))}\right) < 0.$$

The proposition yields a number of nontrivial results given the feedback loops involved in our model. Let's start with optimal land use, namely, with the forest land surface, $f^*$. A bigger regeneration rate $\beta$ raises the level of biodiversity as a direct effect but it does also induce an original second-round effect. Indeed through the agroecological productivity upward shift, it increases the agricultural output for a given agricultural land, $1 - f^*$. This may either lead to increase agricultural land (that to decrease $f^*$) to take advantage of this higher productivity or to decrease it (if the productivity increment is large enough) to rise the forest land, which would generate more biodiversity without penalizing consumption of the agricultural good. Subsequent posterior round effects could occur according to the same opposite mechanisms. A third possibility is that optimal agricultural land remains overall unchanged by the shock on the regeneration rate $\beta$ as the direct and indirect effects may cancel out. Due to our functional specifications (linearity of the agroecological productivity effect and Cobb-Douglas utility function mainly), this is what happens here. Things are different when either $\alpha$ or $\gamma$ goes up. An increase in $\gamma$ increases the weight of the agricultural good in the instantaneous utility function leading to more priority to the consumption of this good, which tends to increase agricultural land (or decrease $f^*$). This is the principal mechanism generating the decline of forest surface: as the elasticity parameter $\gamma$ rises, it eventually dominates indirect effects originating in the subsequent biodiversity decline and the posterior weakening of the agroecological productivity effect. A similar picture occurs with $\alpha$: if this parameter is augmented, the strength of the agroecological productivity boost drops, which reduces the incentives to preserve and/or increase biodiversity, leading to the erosion of forest land. Again in this case, feedback loop effects arise that mitigate the latter effect but they end up dominated.

As to biodiversity consumption, $C_B$, the comparative statics obtained, while quite intuitive, are indeed far from trivial except for $\beta$. A larger regeneration rate of biodiversity, $\beta$, leaves room for more harvesting as it makes biodiversity more abundant. This seems natural and indeed it is so in our analytical frame: by Proposition 11.1, one gets $C_B^* = \frac{(1-\gamma)\beta}{1-\alpha\gamma}(f^*)^\theta$; the comparative static derives automatically from the fact that the optimal forest land $f^*$ is independent of $\beta$, which is, as argued above, due to our chosen functional specifications. This need not be the case in general if the optimal forest land surface does depend on $\beta$. A larger $\beta$ indeed also favors the agroecological productivity effect since this increases directly the biodiversity amount and may further lead to less land for agricultural production if the productivity boost is big enough. Therefore, everything can happen in the general case depending on the function specifications adopted. Equally intuitive is the comparative static with respect to $\gamma$: the larger the elasticity of instantaneous utility with respect to this parameter, more weight is given to the consumption of the agricultural good, which eventually leads to a drop in biodiversity harvesting. More intricate, as the elasticity of the agricultural good production with respect to agricultural land, $\alpha$, goes up, more land will be devoted to agriculture (and more agricultural goods will be consumed). This in turn leads to less biodiversity and subsequent wild good consumption drop. However, as $\alpha$ increases, the fraction of biodiversity produced by Nature and going to optimal bush food consumption, that is, $\frac{1-\gamma}{1-\alpha\gamma}$,[3] goes up. The latter effect dominates the former, and bush food consumption goes actually up when $\alpha$ rises, which is far from trivial. Again as for comparative statics above, these are the dominant effects which reflect the qualitative results obtained in Proposition 11.2. Feedback effects may mitigate the latter but end up dominated.

Let us finish this section with the comparative statics of the biodiversity stock, which is the central variable of our model being the unique dynamic variable, that's the one which drives all the dynamics of the model (given decisions/controls) in the canonical intertemporal problem. Recall that:

$$B^* = \beta\ (f^*)^{\theta} - C_B^* = \beta \frac{\gamma(1-\alpha)}{1-\alpha\gamma}\ (f^*)^{\theta}.$$

An increase in $\beta$ naturally increases the stock of biodiversity despite it also raising wild good consumption. This property is very likely to be robust to changes in the functional specifications of the utility function and the agroecological productivity effect. Another easy property can be found using the results of some of the comparative statics characterized above: since an increase in $\alpha$ decreases the surface of the forest land and raises wild food consumption, this leads to a non-ambiguous fall in the biodiversity stock. Inversely, if the strength of the agroecological productivity boost is raised as $1-\alpha$ increases, the stock of biodiversity increases unambiguously pushed by the two latter mechanisms (operated in opposite direction). The comparative statics of $B^*$ with respect to $\gamma$ is a bit more involved. Indeed, a larger elasticity of utility to the agricultural good decreases both the optimal forest land and consumption of the bush goods, resulting in an ambiguous effect on $B^*$. Proposition 11.2 states that **whatever the parameter values** ($\theta$, $\alpha$, $\beta$ and $\gamma$), the magnitude of the decrease in bush goods consumption is larger than the reduction in biodiversity originating in optimal forest land shrinking. We shall see that such a property cannot hold in the dynamic counterpart of the model, at least not in the long term.

[3] This results from $C_B^* = \frac{(1-\gamma)\beta}{1-\alpha\gamma}\ (f^*)^{\theta}$ in Proposition 11.1.

## 11.4 Exploring the Dynamics

We now come back to our optimal control problem, reformulated with the functional specification choices made in the static problem in the previous section:

$$\max_{f, c_B} \int_0^{\infty} B^{(1-\alpha)\gamma}\ (1-f)^{\alpha\gamma}\ C_B^{1-\gamma}\ e^{-\delta\, t}\ dt \tag{11.11}$$

subject to

$$\dot{B} = \beta\ B^{1-\theta}\ f^{\theta} - C_B,$$

with $B(0) = B_0 > 0$ given.

### 11.4.1 Solving the Optimal Control Problem

The current value Hamiltonian writes as follows:

$$\tilde{H} = B^{(1-\alpha)\gamma}\ (1-f)^{\alpha\gamma}\ C_B^{1-\gamma} + \mu\left(\beta\ B^{1-\theta}\ f^{\theta} - C_B\right),$$

where $\mu$ is the current-valued shadow price of the biodiversity stock, $B$. The corresponding first-order Pontryagin conditions are (with $U \equiv B^{(1-\alpha)\gamma}\ (1-f)^{\alpha\gamma}\ C_B^{1-\gamma}$ is the instantaneous utility):

$$(1-\gamma)\ \frac{U}{C_B} = \mu, \tag{11.12}$$

$$\alpha\gamma \frac{U}{1-f} = \mu\theta\beta B^{1-\theta}\ f^{\theta-1} \tag{11.13}$$

$$(1-\alpha)\gamma \frac{U}{B} + \mu(1-\theta)\beta B^{-\theta}\ f^{\theta} = -\dot{\mu} + \delta\mu, \tag{11.14}$$

with the transversality condition: $\lim_{t\to+\infty} e^{-\rho t}\mu B = 0$. Equation (11.12) gives the necessary optimality condition for $C_B$: it equalizes the marginal utility from one additional unit of biodiversity harvesting to biodiversity shadow price, $\mu$. Equation (11.13) is the necessary optimality condition for $f$: the marginal disutility from a one additional unit of forest land (left-hand side of the equation) should be equal to the marginal benefit in terms of biodiversity increment evaluated at the biodiversity shadow price. Equation (11.14) is the co-state equation, which delivers the time variation of the (current-valued) shadow price of biodiversity resulting from time discounting (term $\delta\mu$ in the right-hand side), the marginal utility gained from one additional unit of biodiversity (through the agroecological productivity effect leading to more consumption of the agricultural good), and the benefit from increasing the biodiversity stock evaluated at the corresponding shadow price. The transversality condition is standard; it stated that the marginal discounted value of biodiversity evaluated at the current-valued shadow price should be exhausted asymptotically. The second-order (sufficiency) optimality conditions, being quite intricate, are proved and reported in the Appendix.

Before we study the existence of stationary states for the system of necessary optimality conditions uncovered above, we first define a stationary state accurately.

**Definition 11.1** The 4-uple $\{f^s, C_B^s, B^s, \mu^s\} \in \mathbb{R}_+^4$ with $f^s < 1$ is a stationary state of the optimal control problem (11.11) if it solves the system of Eqs. (11.3), (11.12), (11.13), and (11.14), plus the transversality condition, under the stationarity conditions: $\dot{B} = \dot{\mu} = 0$.

The corresponding system of equations solving for stationary states is therefore given by Eqs. (11.12)–(11.13) plus the stationarized state and co-state equations:

$$C_B^s = \beta \left(B^s\right)^{1-\theta} \left(f^s\right)^{\theta} \tag{11.15}$$

$$(1-\alpha)\gamma \frac{U^s}{B^s} + \mu^s (1-\theta)\beta \left(B^s\right)^{-\theta} \left(f^s\right)^{\theta} = \delta\mu^s, \tag{11.16}$$

where $U^s$ stands for instantaneous utility evaluated at the steady state. Notice that if such a state exists, the transversality conditions is automatically checked. The next proposition proves existence and uniqueness of such a stationary state and explores local stability.

**Proposition 11.3** *For the 4-uple* $\{f^s, C_B^s, B^s, \mu^s\}$, *check the following properties:*

- *It exists and is unique. It's given by:*

$$f^s = \frac{(1-\gamma)\theta}{\alpha\gamma + (1-\gamma)\theta},$$

$$B^s = \left[\frac{\beta(1-\theta+\gamma(\theta-\alpha))}{\delta(1-\gamma)}\right]^{\frac{1}{\theta}} f^s,$$

$$C_B^s = \beta \left(B^s\right)^{1-\theta} \left(f^s\right)^{\theta},$$

*and*

$$\mu^s = (1-\alpha)\frac{U^s}{C_B^s}.$$

- *The stationary state is saddle point stable.*

We skip the algebraic details of the computation of the steady state; it's easy. A remarkable property of the dynamic problem at steady state with respect to the static counterpart is the role of the time discounting parameter (that's the so-called impatience rate), $\delta$, which of course does not show up in static settings. Remarkably enough, with our canonical specifications, **the stock of biodiversity converges to 0 when $\delta$ becomes infinitely large. Indeed, the whole economy goes to extinction as the vanishing stock of biodiversity also leads to the collapse of the agricultural sector in our case.** That's what also makes this model useful as a benchmark frame.

We develop now the analysis of the stationary state local stability before moving to the economic exploration. To study the local stability, we need to write the dynamics. Using that $\mu$ appears in both Eqs. 11.12 and 11.13,

$$C_B = \chi\beta B^{1-\theta}(1-f)\, f^{\theta-1}, \tag{11.17}$$

where $\chi = \frac{(1-\gamma)\theta}{\alpha\gamma}$. As a consequence, the state dynamics is given as follows.

$$\dot{B} = \beta\, B^{1-\theta} f^{\theta}\left(1-\chi\frac{1-f}{f}\right). \tag{11.18}$$

Moreover, using Eq. 11.12 in Eq. 11.14 yields

$$(1-\alpha)\gamma\frac{\chi\beta B^{-\theta}(1-f)\, f^{\theta-1}}{(1-\gamma)} + (1-\theta)\beta B^{-\theta} f^{\theta} = -\frac{\dot{\mu}}{\mu} + \delta,$$

which can be written as

$$\frac{\dot{\mu}}{\mu} = \delta - \beta B^{-\theta} f^{\theta}\left((1-\alpha)\frac{\theta}{\alpha}\frac{1-f}{f} + (1-\theta)\right).$$

As, according to Eq. 11.12 it can be obtained

$$(1-\gamma)(\chi\beta)^{-\gamma}\; B^{(\theta-\alpha)\gamma}\,(1-f)^{(\alpha-1)\gamma}\, f^{(1-\theta)\gamma} = \mu.$$

As a consequence,

$$(\theta-\alpha)\,\gamma\frac{\dot{B}}{B} - (\alpha-1)\,\gamma\frac{\dot{f}}{1-f} + (1-\theta)\,\gamma\frac{\dot{f}}{f} = \frac{\dot{\mu}}{\mu},$$

from which we can write the dynamics of $f$.

$$\left(-(\alpha-1)\frac{1}{1-f}+(1-\theta)\frac{1}{f}\right)\gamma\dot{f}$$

$$=\delta-\beta B^{-\theta}f^{\theta}\begin{pmatrix}\frac{\theta}{\alpha}\left((1-\alpha)-(\theta-\alpha)(1-\gamma)\right)\frac{1-f}{f}\\+(1-\theta)+(\theta-\alpha)\gamma\end{pmatrix} \tag{11.19}$$

$$=\delta-\beta B^{-\theta}f^{\theta}\left((1-\theta)+(\theta-\alpha)\gamma\right)\left(\frac{\theta}{\alpha}\frac{1-f}{f}+1\right). \tag{11.20}$$

It can be noticed that $\frac{(1-\theta)}{f}-\frac{(\alpha-1)}{1-f}\neq 0$, as $f\neq\frac{\theta-1}{\theta-\alpha}$.
The Jacobian matrix of system 11.18–11.20 at steady state is given as follows:

$$J=\begin{bmatrix}0 & \frac{\beta\eta^{\frac{1-\theta}{\theta}}\chi}{f^s}\\ \frac{\beta\eta^{-(1/\theta)}\delta f^s}{\alpha\gamma\left(\frac{(1-\theta)}{f}-\frac{(\alpha-1)}{1-f}\right)} & \frac{\delta((\alpha-\theta)\gamma+\theta(1-\alpha))}{\alpha f^s\gamma\left(\frac{(1-\theta)}{f}-\frac{(\alpha-1)}{1-f}\right)}\end{bmatrix}$$

where $\frac{\beta(1-\theta+\gamma(\theta-\alpha))}{\delta(1-\gamma)}=\eta$. The eigenvalues are the solution of $\Delta(\lambda)=0$, where the characteristic polynomial is given by:

$$\Delta(\lambda)=\lambda^2-tr(J)\lambda+\det(J),$$

where $\det(J)>0$ and $\mathrm{tr}(J)>0$ are respectively determinant and trace of $J$. As a consequence, eigenvalues are real, with opposite sign. So the stationary state is saddle point stable. After validating the optimality and (local) stability of the steady state, we explore here below its economic outcomes and compare them with those of the static counterpart.

### *11.4.2 Economic Properties*

Before getting to the comparison task, we start with the comparative statics of the stationary state. As we shall see, they are more involved than in the static counterpart.

**Proposition 11.4** *The following comparative statics hold:*

1. $\frac{\partial f^*}{\partial\beta}=0$, $\frac{\partial f^*}{\partial\alpha}<0$, $\frac{\partial f^*}{\partial\gamma}<0$, *and* $\frac{\partial f^*}{\partial\delta}=0$.
2. *It follows that* $\frac{\partial B^*}{\partial\beta}>0$, $\frac{\partial B^*}{\partial\alpha}<0$, *and* $\frac{\partial B^*}{\partial\delta}<0$.
3. *The sign of* $\frac{\partial B^*}{\partial\gamma}$ *is ambiguous; it depends on the parameters' values. Ceteris paribus,* $B^*$ *increases with* $\gamma$ *if the strength of the agroecological productivity*

*effect is strong enough (i.e., $\alpha$ is small enough), and it's a decreasing function of $\gamma$ when the latter productivity effect is weak enough (i.e., $\alpha$ is large enough).*

4. $\frac{\partial C_B^*}{\partial \beta} > 0$, $\frac{\partial C_B^*}{\partial \alpha} < 0$, *and* $\frac{\partial C_B^*}{\partial \delta} < 0$.
5. $\frac{\partial C_B^*}{\partial \gamma}$ *is ambiguous; it depends on the parameters' values. Similarly to* $B^*$, $C_B^*$ *increases with* $\gamma$ *if* $\alpha$ *is small enough, and it's a decreasing function of* $\gamma$ *when* $\alpha$ *is large enough.*

A few observations are worth doing before discussing the economic mechanisms involved in these comparative statics. First of all, and as mentioned above, the stationary states of dynamic systems do depend on the way the optimizer/planner values future utility flows compared to the current ones, that's on the time discounting rate, $\delta$. A very large time discounting brings the whole economy close to extinction asymptotically as demonstrated in the previous section. A larger time discounting reduces the stock of biodiversity and the consumption of the bush food.[4] Second, some comparative results are reversed with respect to the static counterpart. This is clear in the fourth comparative static: bush food consumption is a decreasing function of the strength of the agroecological productivity effect in the optimal static solution (see Proposition 11.1), and it's increasing the dynamic case at the stationary equilibrium. Third, and more frequently, a new type of results with respect to the static case may emerge as the dynamic interaction of direct and feedback effects may lead to much more complex pictures in the dynamic settings, even at the steady state, yielding non-monotonic relationships and the like. In our case, as featured in comparative statics 3 and 5, here we enhance the role of the parameter (that's $\alpha$) which somehow measures, as repeatedly invoked above, the strength of the feedback mechanism from the amount of biodiversity to the productivity of the agricultural sector.

We shall concentrate on the comparative statics 3 to 5, those which differ from those of the static case, precisely those with respect to the key parameters $\alpha$ and $\gamma$. We skip most of the algebraic details. Let's start with the stock variable, $B^s$ (comparative static 3) Recall that it's given by

$$B^s = \left[\frac{\beta(1-\theta+\gamma\,(\theta-\alpha))}{\delta\,(1-\gamma)}\right]^{\frac{1}{\theta}} f^s.$$

It follows that

$$\frac{\partial B^s}{\partial \alpha} = \left[\frac{\beta(1-\theta+\gamma\,(\theta-\alpha))}{\delta\,(1-\gamma)}\right]^{\frac{1}{\theta}-1}$$

[4] It could be also shown that the same property holds for the consumption of the agricultural good due to the agricultural sector productivity feedback effect of falling biodiversity similarly to the extinction story told in the previous section. We skip computations for lack of space.

$$\left(\frac{\partial f^s}{\partial \alpha}\left[\frac{\beta(1-\theta+\gamma(\theta-\alpha))}{\delta(1-\gamma)}\right]-f^s\frac{\beta\gamma}{\theta\delta(1-\gamma)}\right)<0,$$

and

$$\begin{aligned}\frac{\frac{\partial B^s}{\partial \gamma}}{B^s}&=\frac{1}{\theta}\frac{\frac{\beta(1-\alpha)}{\delta(1-\gamma)^2}}{\frac{\beta(1-\theta+\gamma(\theta-\alpha))}{\delta(1-\gamma)}}+\frac{\frac{\partial f^s}{\partial \gamma}}{f^s}\\&=\frac{1}{(1-\gamma)}\left(\frac{1}{\theta}\frac{(1-\alpha)}{1-\theta+\gamma(\theta-\alpha)}-\frac{\alpha}{(\alpha-\theta)\gamma+\theta}\right).\end{aligned}$$

Let $z=(\alpha-\theta)\gamma+\theta$

$$\frac{\frac{\partial B^s}{\partial \gamma}}{B^s}=\frac{1}{\theta(1-\gamma)}\left(\frac{z(1-\alpha)-\alpha\theta(1-z)}{(1-z)z}\right).$$

The sign of $\frac{\partial B^s}{\partial \gamma}$ is the same as the sign of $z(1-\alpha)-\alpha\theta(1-z)$, which is ambiguous. For $\alpha=0.9, \theta=0.5, \gamma=0.1$, $\frac{\partial B^s}{\partial \gamma}<0$, whereas for $\alpha=0.1, \theta=0.5, \gamma=0.1$, $\frac{\partial B^s}{\partial \gamma}>0$. It's easy to show that the derivative $\frac{\partial B^s}{\partial \gamma}$ is decreasing with respect to $\alpha$, given the magnitude of $\alpha$ and $\gamma$, both strictly lower than 1. An even more direct way to see this property is to study the limit cases of $\frac{\partial B^s}{\partial \gamma}$ when $\alpha$ goes respectively to 1 and 0.

We finally turn to $C_B^s$. As $C_B^s=\beta(B^s)^{1-\theta}(f^s)^{\theta}$, we get

$$\begin{aligned}\frac{\partial C_B^s/\partial\alpha}{C_B^s}&=(1-\theta)\left(\frac{\partial B^s/\partial\alpha}{B^s}\right)+\theta\left(\frac{\partial f^s/\partial\alpha}{f^s}\right)<0\\\frac{\partial C_B^s/\partial\beta}{C_B^s}&=(1-\theta)\left(\frac{\partial B^s/\partial\beta}{B^s}\right)>0\\\frac{\partial C_B^s/\partial\gamma}{C_B^s}&=(1-\theta)\left(\frac{\partial B^s/\partial\gamma}{B^s}\right)+\theta\left(\frac{\partial f^s/\partial\gamma}{f^s}\right)\\&=(1-\theta)\left(\frac{\partial B^s/\partial\gamma}{B^s}\right)-\frac{\alpha\theta}{(1-\gamma)((\alpha-\theta)\gamma+\theta)}.\end{aligned}$$

We can observe that:

$$\begin{aligned}\frac{\partial C_B^s/\partial\gamma}{C_B^s}&=\frac{1}{(1-\gamma)z}\left[\frac{(1-\theta)}{\theta}\left(\frac{z(1-\alpha)-\alpha\theta(1-z)}{(1-z)}\right)-\alpha\theta\right]\\&=\frac{1}{\theta(1-z)(1-\gamma)z}((1-\theta)z(1-\alpha)-\alpha\theta(1-z)).\end{aligned}$$

However, the sign of $(1-\theta)\,z\,(1-\alpha) - \alpha\theta\,(1-z)$ is ambiguous. Indeed, $\alpha = 0.9, \theta = 0.5, \gamma = 0.9$, then $a < 0$ but for $\alpha = 0.1$, then $a > 0$. Similarly to the comparative static of $B^s$, we obtain property 5 in Proposition 5.

Let us now comment on the economic mechanisms involved and compare with the static counterpart outcomes. First of all, we note that the property that the biodiversity stock increases with the strength of the agroecological productivity effect (or decreases with $\alpha$) also holds in the dynamic case at the stationary equilibrium. In the static case, this relationship is clear-cut: a drop in $\alpha$ raises the surface of the forest land and decreases wild food consumption, which eventually yields a non-ambiguous (and potentially strong) increment in the biodiversity stock. While we get the same corresponding qualitative relationship in the dynamic case (which is indeed an indicator of the well-posedness of our problem), the mechanisms are not the same. As stated in property 4 of Proposition 11.4, the impact of a decrease in $\alpha$ on the biodiversity stock is much more involved in the dynamic case: while the positive forest land effect is still present, consumption of bush meat goes in the opposite direction—it increases when $\alpha$ drops. However, the former positive effect dominates for any values of the parameters of the model. The fact that the comparative static for $C_B^s$ with respect to $\alpha$ is reversed in the dynamic case at the stationary equilibrium is not surprising per se considering the richness of feedback effects in our model, not speaking about the forces governing convergence to the steady state (including time discounting): while a rising stock of biodiversity will push upward its harvesting in first place, it also rises the incentives to produce the agricultural good due to the agroecological productivity effect, which goes in opposite direction relatively to the first effect, not speaking about the dynamics of convergence to the steady state which in all cases alter the size of the different involved effects over time, especially in the medium and long run.

Let's move now to the comparative static of $B^s$ with respect to $\gamma$; that's comparative static 3.: it's similar to the comparative static 5 concerning $C_B^s$ with respect to the same parameter. We shall concentrate on the $B^s$; similar arguments can be used for $C_B^s$ using the relationship relating both variables and $f^s$ given in Proposition 11.3. As explained in Sect. 11.3.2, the comparative statics of $B^*$ with respect to $\gamma$ are a priori intricate even in the static case: in such a case, an agricultural economy which exhibits a larger elasticity of utility to the agricultural good, that's with a stronger preference to the latter good, would initially decrease both the optimal forest land and consumption of the bush goods, resulting in an ambiguous direct effect on $B^*$. However in the static case, we have found that **whatever the parameter values** ($\theta$, $\alpha$, $\beta$ and $\gamma$), the magnitude of the decrease in bush goods consumption is larger than the reduction in biodiversity originating in optimal forest land shrinking, accounting for all the feedback effects generated. In the dynamic case at the stationary equilibrium, everything depends on the parameters' values. To simplify our discussion, we focus on the key parameter in the genesis of feedback effects, $\alpha$. Proposition 11.4 shows that when the agroecological productivity effect is strong enough ($\alpha$ small enough), we get the static picture, and we get the reverse in the opposite case. This is rather intuitive: in a dynamic model, the feedback effects play longer over time (though constrained by the convergence to steady state forces as time advances); this leads to a bigger role for the forest land surface effect, which is inherent in the agroecological effect, compared to the one channeled through bush food consumption. This yields in particular to the (long-term) property that as the agricultural society preferences move away from bush food consumption ($\gamma$ increases), the stock of biodiversity is raised only if the agroecological productivity effect is strong enough, which is a quite interesting result.

## 11.5 Concluding Remarks

We have shown both in the static and dynamic cases how the feedback effect generated by the agroecological productivity boost shapes the qualitative properties within a canonical agricultural model with land use control and consumption (optimal) arbitrage between agricultural and wild goods. The interaction between the feedback effect and the mechanisms inherent in intertemporal optimization has been shown to deliver several highly interesting results (for the long-term equilibrium), with some policy relevance. Several methodological lessons have been drawn along the way.

We believe that this model is a useful benchmark, and we have explained why in several places in the main text. The fact remains that it's a benchmark. Several extensions are worth doing starting with the analysis (possibly numerical) of the general model with nonlinear feedback effects, for example. Also, more general consumption preferences are interesting to incorporate, possibly with endogenous cultural dynamics moving societies away from wild food. Last but not least, more convincing specifications of biodiversity dynamics are needed starting with the endogenization of the regeneration rate.

## Appendix: Sufficiency Analysis of the Optimal Control Problem

We now study the sufficient optimality conditions. We consider $\tilde{H}$ as a function of $B$ and $f$. Indeed, $\tilde{H} = B^{(1-\alpha)\gamma}\,(1-f)^{\alpha\gamma}\,C_B^{1-\gamma} + \mu\left(\beta\,B^{1-\theta} f^{\theta} - C_B\right)$. According to Eq. 11.17,

$$\tilde{H} = B^{(1-\alpha)\gamma+(1-\theta)(1-\gamma)}\,(1-f)^{\alpha\gamma+1-\gamma}\,(\chi\beta)^{1-\gamma}\,f^{(\theta-1)(1-\gamma)}$$
$$+\,\mu\beta\,B^{1-\theta} f^{\theta-1}\,(f-\chi\,(1-f))\,.$$

Moreover, according to Eq. 11.12

$$(1-\gamma)\,B^{(1-\alpha)\gamma}\,(1-f)^{\alpha\gamma}\,C_B^{-\gamma} = \mu.$$

Using 11.17, it can be written as

$$(\chi\beta)^{-\gamma}(1-\gamma)\,B^{(1-\alpha)\gamma-\gamma(1-\theta)}\,(1-f)^{(\alpha-1)\gamma}\,f^{\gamma(1-\theta)}=\mu,$$

$$\tilde{H}=\frac{1}{\theta}(\chi\beta)^{1-\gamma}\,B^{(1-\alpha)\gamma+(1-\gamma)(1-\theta)}f^{(\theta-1)(1-\gamma)}(1-f)^{(\alpha-1)\gamma}\gamma\,[\theta+(\alpha-\theta)f].$$

As $\theta+(\alpha-\theta)f=\theta(1-f)+\alpha f>0$, letting $h=\frac{\gamma}{\theta}(\chi\beta)^{1-\gamma}$,

$$\tilde{H}=hB^{1-\theta+\gamma(\theta-\alpha)}f^{(\theta-1)(1-\gamma)}(1-f)^{(\alpha-1)\gamma}(\theta+f(\alpha-\theta))>0.$$

In order to compute the Hessian Matrix of $\tilde{H}$, let us notice that

$$\frac{\partial\tilde{H}}{\partial B}=(1-\theta+\gamma(\theta-\alpha))\frac{\tilde{H}}{B}$$

$$\frac{\partial^2\tilde{H}}{\partial B^2}=(1-\theta+\gamma(\theta-\alpha))(-\theta+\gamma(\theta-\alpha))\frac{\tilde{H}}{B^2}$$

$$=-(1-\theta+\gamma(\theta-\alpha))((1-\gamma)\theta+\alpha\gamma)\frac{\tilde{H}}{B^2}.$$

As $1-\theta+\gamma(\theta-\alpha)=(1-\alpha)\gamma+(1-\theta)(1-\gamma)$, then $1-\theta+\gamma(\theta-\alpha)>0$. Thus, as a consequence,

$$\frac{\partial^2\tilde{H}}{\partial B^2}<0.$$

Moreover,

$$\frac{\partial\tilde{H}}{\partial B\partial f}=(1-\theta+\gamma(\theta-\alpha))\frac{\partial\tilde{H}/\partial f}{B}$$

and

$$\frac{\partial\tilde{H}}{\tilde{H}\partial f}=\frac{(\theta-1)(1-\gamma)}{f}+\frac{(1-\alpha)\gamma}{1-f}+\frac{\alpha-\theta}{\theta+f(\alpha-\theta)}$$

$$=\frac{-(1-\theta+f(\theta-\alpha))(-(1-f)\theta(1-\gamma)+f\alpha\gamma)}{f(1-f)(\theta+f(\alpha-\theta))}.$$

Let $g(f)=\frac{-(1-\theta+f(\theta-\alpha))(-(1-f)\theta(1-\gamma)+f\alpha\gamma)}{f(1-f)(\theta+f(\alpha-\theta))}$

$$\frac{\partial^2\tilde{H}}{\partial f^2}=\left(\frac{\partial\tilde{H}}{\partial f}\right)g(f)+\tilde{H}\left(\frac{(1-\theta)(1-\gamma)}{f^2}+\frac{(1-\alpha)\gamma}{(1-f)^2}-\frac{(\alpha-\theta)^2}{(\theta+f(\alpha-\theta))^2}\right).$$

The Hessian matrix of $\tilde{H}$ as a function of $B$ and $\mu$ is given as follows:

$$\mathcal{H}(B, f) = \begin{bmatrix} -(1-\theta+\gamma(\theta-\alpha))((1-\gamma)\theta+\alpha\gamma)\frac{\tilde{H}}{B^2} & (1-\theta+\gamma(\theta-\alpha))\frac{\partial\tilde{H}/\partial f}{B} \\ (1-\theta+\gamma(\theta-\alpha))\frac{\partial\tilde{H}/\partial f}{B} & \frac{\partial^2\tilde{H}}{\partial f^2} \end{bmatrix}$$

$$\det(\mathcal{H}(B, f)) = (1-\theta+\gamma(\theta-\alpha))\frac{\tilde{H}^2}{B^2}\left[-((1-\gamma)\theta+\alpha\gamma)g'(f)-g^2(f)\right].$$

Moreover, in the neighborhood of the steady state $f^s = \frac{(1-\gamma)\theta}{\alpha\gamma+(1-\gamma)\theta}$, $g(f) \simeq 0$ and $g'(f) < 0$. As $(1-\theta+\gamma(\theta-\alpha)) > 0$, then $\det(\mathcal{H}(B, f)) > 0$. Moreover, as $\frac{\partial^2\tilde{H}}{\partial B^2} < 0$, then the optimum is a maximum.

**Acknowledgements** Raouf Boucekkine acknowledges the financial support from the French government under the "France 2030" investment plan managed by the French National Research Agency Grant ANR-17-EURE0020.

## References

Baier, C., Gross, A., Thevs, N., & Glaser, B. (2023). Effects of agroforestry on grain yield of maize (zea mays l.)—A global meta-analysis. *Frontiers in Sustainable Food Systems, 7*, 1167686.

Beillouin, D., Ben-Ari, T., Malézieux, E., Seufert, V., & Makowski, D. (2021). Positive but variable effects of crop diversification on biodiversity and ecosystem services. *Global Change Biology, 27*(19), 4697–4710.

Bommarco, R., Kleijn, D., & Potts, S. G. (2013). Ecological intensification: harnessing ecosystem services for food security. *Trends in Ecology & Evolution, 28*(4), 230–238.

Centeno-Alvarado, D., Lopes, A. V., & Arnan, X. (2023). Fostering pollination through agroforestry: A global review. *Agriculture, Ecosystems & Environment, 351*, 108478.

Chaudhary, A., Burivalova, Z., Koh, L. P., & Hellweg, S. (2016). Impact of forest management on species richness: Global meta-analysis and economic trade-offs. *Scientific Reports, 6*(1), 23954.

Dainese, M., Martin, E. A., Aizen, M. A., Albrecht, M., Bartomeus, I., Bommarco, R., Carvalheiro, L. G., Chaplin-Kramer, R., Gagic, V., Garibaldi, L. A., et al. (2019). A global synthesis reveals biodiversity-mediated benefits for crop production. *Science Advances, 5*(10), eaax0121.

Duru, M., Therond, O., Martin, G., Martin-Clouaire, R., Magne, M.-A., Justes, E., Journet, E.-P., Aubertot, J.-N., Savary, S., Bergez, J.-E., et al. (2015). How to implement biodiversity-based agriculture to enhance ecosystem services: A review. *Agronomy for Sustainable Development, 35*, 1259–1281.

Foley, J. A., DeFries, R., Asner, G. P., Barford, C., Bonan, G., Carpenter, S. R., Chapin, F. S., Coe, M. T., Daily, G. C., Gibbs, H. K., et al. (2005). Global consequences of land use. *Science, 309*(5734), 570–574.

Garibaldi, L. A., Gemmill-Herren, B., D'Annolfo, R., Graeub, B. E., Cunningham, S. A., & Breeze, T. D. (2017). Farming approaches for greater biodiversity, livelihoods, and food security. *Trends in Ecology & Evolution, 32*(1), 68–80.

Kremen, C., Iles, A., and Bacon C. (2012). Diversified farming systems: an agroecological, systems-based alternative to modern industrial agriculture. *Ecology and Society, 17*(4), 44. https://dx.doi.org/10.5751/ES-05103-170444

Kuyah, S., Whitney, C. W., Jonsson, M., Sileshi, G. W., Öborn, I., Muthuri, C. W., & Luedeling, E. (2019). Agroforestry delivers a win-win solution for ecosystem services in sub-saharan africa. a meta-analysis. *Agronomy for Sustainable Development, 39*, 1–18.

Nair, P. R. (2014). Grand challenges in agroecology and land use systems. *Frontiers in Environmental Science, 2*, 1.

Nair, P. R., Kumar, B. M., & Nair, V. D. (2021). *An introduction to agroforestry: Four decades of scientific developments*. Springer.

Newbold, T., Hudson, L. N., Hill, S. L., Contu, S., Lysenko, I., Senior, R. A., Börger, L., Bennett, D. J., Choimes, A., Collen, B., et al. (2015). Global effects of land use on local terrestrial biodiversity. *Nature, 520*(7545), 45–50.

Ortiz, A. M. D., Outhwaite, C. L., Dalin, C., & Newbold, T. (2021). A review of the interactions between biodiversity, agriculture, climate change, and international trade: research and policy priorities. *One Earth, 4*(1), 88–101.

Potapov, P., Hansen, M. C., Pickens, A., Hernandez-Serna, A., Tyukavina, A., Turubanova, S., Zalles, V., Li, X., Khan, A., Stolle, F., et al. (2022a). The global 2000-2020 land cover and land use change dataset derived from the landsat archive: first results. *Frontiers in Remote Sensing, 3*, 856903.

Potapov, P., Turubanova, S., Hansen, M. C., Tyukavina, A., Zalles, V., Khan, A., Song, X.-P., Pickens, A., Shen, Q., & Cortez, J. (2022b). Global maps of cropland extent and change show accelerated cropland expansion in the twenty-first century. *Nature Food, 3*(1), 19–28.

Pretty, J. (2008). Agricultural sustainability: Concepts, principles and evidence. *Philosophical Transactions of the Royal Society B: Biological Sciences, 363*(1491), 447–465.

Pumariño, L., Sileshi, G. W., Gripenberg, S., Kaartinen, R., Barrios, E., Muchane, M. N., Midega, C., & Jonsson, M. (2015). Effects of agroforestry on pest, disease and weed control: A meta-analysis. *Basic and Applied Ecology, 16*(7), 573–582.

Shukla, P. R., Skea, J., Calvo Buendia, E., Masson-Delmotte, V., Pörtner, H. O., Roberts, D., Zhai, P., Slade, R., Connors, S., Van Diemen, R., et al. (2019). IPCC, 2019: Climate change and land: an ipcc special report on climate change, desertification, land degradation, sustainable land management, food security, and greenhouse gas fluxes in terrestrial ecosystems.

Staton, T., Walters, R. J., Smith, J., & Girling, R. D. (2019). Evaluating the effects of integrating trees into temperate arable systems on pest control and pollination. *Agricultural Systems, 176*, 102676.

Tittonell, P., Klerkx, L., Baudron, F., Félix, G. F., Ruggia, A., Van Apeldoorn, D., Dogliotti, S., Mapfumo, P., & Rossing, W. A. (2016). Ecological intensification: local innovation to address global challenges. *Sustainable Agriculture Reviews, 19*, 1–34.

Torralba, M., Fagerholm, N., Burgess, P. J., Moreno, G., & Plieninger, T. (2016). Do European agroforestry systems enhance biodiversity and ecosystem services? A meta-analysis. *Agriculture, Ecosystems & Environment, 230*, 150–161.

Tscharntke, T., Clough, Y., Wanger, T. C., Jackson, L., Motzke, I., Perfecto, I., Vandermeer, J., & Whitbread, A. (2012). Global food security, biodiversity conservation and the future of agricultural intensification. *Biological Conservation, 151*(1), 53–59.

Udawatta, R. P., Rankoth, L. M., & Jose, S. (2019). Agroforestry and biodiversity. *Sustainability, 11*(10), 2879.

van Noordwijk, M., Duguma, L. A., Dewi, S., Leimona, B., Catacutan, D. C., Lusiana, B., Öborn, I., Hairiah, K., & Minang, P. A. (2018). SDG synergy between agriculture and forestry in the food, energy, water and income nexus: Reinventing agroforestry? *Current Opinion in Environmental Sustainability, 34*, 33–42.

Waldron, A., Garrity, D., Malhi, Y., Girardin, C., Miller, D., & Seddon, N. (2017). Agroforestry can enhance food security while meeting other sustainable development goals. *Tropical Conservation Science, 10*, 1940082917720667.

Wezel, A., Casagrande, M., Celette, F., Vian, J.-F., Ferrer, A., & Peigné, J. (2014). Agroecological practices for sustainable agriculture. A review. *Agronomy for Sustainable Development, 34*(1), 1–20.

Winkler, K., Fuchs, R., Rounsevell, M., & Herold, M. (2021). Global land use changes are four times greater than previously estimated. *Nature Communications, 12*(1), 2501.

Zhang, W., Ricketts, T. H., Kremen, C., Carney, K., & Swinton, S. M. (2007). Ecosystem services and dis-services to agriculture. *Ecological Economics, 64*(2), 253–260.

Zomer, R. J., Neufeldt, H., Xu, J., Ahrends, A., Bossio, D., Trabucco, A., Van Noordwijk, M., & Wang, M. (2016). Global tree cover and biomass carbon on agricultural land: The contribution of agroforestry to global and national carbon budgets. *Scientific Reports, 6*(1), 29987.

Zomer, R. J., Trabucco, A., Coe, R., Place, F., Van Noordwijk, M., & Xu, J. (2014). Trees on farms: An update and reanalysis of agroforestry's global extent and socio-ecological characteristics. Technical report.

# Chapter 12
# Open-Loop Control-Based Linear-Quadratic Stochastic Game with Application to Counter Terror: Farsighted Versus Myopic Policies

**Konstantin Kogan and Dmitry Tsadikovich**

**Abstract** Stochastic linear-quadratic problems are frequently found in optimal control and differential game applications. The typical solution to these problems is based on feedback control, which implies that the state dynamics are observable despite their stochastic nature. In real life, however, this is rarely the case. In this chapter, we overcome this unobservability drawback by deriving an open-loop equilibrium control for a linear-quadratic dynamic game with applications to counter-terror activities characterized by stochastic terrorist resource stocks. We derive an open-loop Nash equilibrium solution and its time-dependent feedback representation, which is based on expected terrorist resources rather than on the true state of the resource stock. We contrast the found equilibrium control with myopic behavior in response to resource dynamics by one or both parties and show that a farsighted party always has an advantage over a myopic party (i) under simultaneous commitments and (ii) when a farsighted party's leader openly commits to actions and the myopic party is a follower responding to the farsighted leader's actions. Furthermore, uncertainty improves the position of the farsighted party in terms of resource goals. In particular, the greater the resource-related uncertainty, the stronger the resource accumulation when terrorists are farsighted and the government is myopic. Uncertainty is detrimental to the government; it increases the economic damage inflicted by the terrorist organization and thus decreases the cost efficiency of the government's counter-terror efforts.

Stochastic linear-quadratic problems are frequently found in optimal control and differential game applications. The typical solution to these problems is based on feedback control, which implies that the state dynamics are observable despite their stochastic nature. In real life, however, this is rarely the case. In this chapter,

K. Kogan (✉) · D. Tsadikovich
Department of Management, Bar-Ilan University, Ramat-Gan, Israel
e-mail: Konstantin.Kogan@biu.ac.il

 
F. El Ouardighi, G. Feichtinger (eds.), *The Unaffordable Price of Static Decision-making Models*, International Series in Operations Research & Management Science 365, https://doi.org/10.1007/978-3-031-88638-6_12

we overcome this unobservability drawback by deriving an open-loop equilibrium control for a linear-quadratic dynamic game with applications to counter-terror activities characterized by stochastic terrorist resource stocks. We derive an open-loop Nash equilibrium solution and its time-dependent feedback representation, which is based on expected terrorist resources rather than on the true state of the resource stock. We contrast the found equilibrium control with myopic behavior in response to resource dynamics by one or both parties and show that a farsighted party always has an advantage over a myopic party (i) under simultaneous commitments and (ii) when a farsighted party's leader openly commits to actions and the myopic party is a follower responding to the farsighted leader's actions. Furthermore, uncertainty improves the position of the farsighted party in terms of resource goals. In particular, the greater the resource-related uncertainty, the stronger the resource accumulation when terrorists are farsighted and the government is myopic. Uncertainty is detrimental to the government; it increases the economic damage inflicted by the terrorist organization and thus decreases the cost efficiency of the government's counter-terror efforts.

## 12.1 Introduction

Stochastic linear-quadratic problems characterized by expected quadratic cost functional and linear stochastic state dynamics have been extensively studied and are used in various applications (e.g., Bismut (1976), Chen and Yong (2001), Tang (2020), Li et al. (2022)). The typical approach used to solve these problems assumes that the system states are observable despite their stochastic nature. Dynamic programming can then be applied that, in the case of a continuous-time formulation, results in the Hamilton-Jacobi-Bellman (HJB) equation or in a system of HJB equations when the problem is reduced to a differential game (e.g., Yong and Zhou (1999), Sun and Yong (2020)). Often, however, the system states are not observable, especially when the environment is stochastic. In those cases, an open-loop approach is more realistic and more practical. Open-loop solvability is not a given in linear-quadratic optimal-control problems. For example, Wei et al. (2019) characterize an open-loop solution of linear-quadratic optimal-control problems with operator coefficients by means of a system of linear coupled forward-backward stochastic differential equations. Kogan and Chernonog (2019) present a numerical approach for a linear-quadratic type of stochastic differential game involving industry-stock-driven competition and contrast the outcome with outcomes from a feedback solution.

The present study derives an analytical, open-loop Nash solution for a linear-quadratic stochastic control formulation with application to a counter-terror game. Differential games have been widely used to model interactions between terrorist and government organizations (e.g., Behrens et al. (2007), Feichtinger and Novak (2008), Zhuang et al. (2010), Crettez and Hayek (2014)). The common approach to tackling these games is to assume observability of the terrorist's resource stocks to derive a feedback policy for the stochastic change in the stock level. In real life, however, the terrorist's resources are not observable. The government does not have accurate information about their resources, and even the terrorist organization can only guess about its total resource stock because of the multiple sources of financing typically involved, its multiple activities, and related uncertainty. This disadvantage has motivated various models intended to estimate the likely number of terrorist threats using queuing theory (e.g., Kaplan (2010, 2015), Seidl et al. (2016)).

In this work, we consider the stochastic counter-terror differential game formulated in Sun et al. (2018) and Li et al. (2021) to determine an open-loop equilibrium instead of attempting to estimate the non-observable level of terrorist threat in a two-party system composed of a government and a terrorist organization. We pay special attention to the effect of different types of myopic behavior on equilibrium policies. We show that an open-loop equilibrium control for the two conflicting parties is characterized by a non-linear function of time and identifies a feedback representation of the equilibrium control that is based on the expected stock of the terrorist resources rather than on the true stock of resources. Furthermore, when one party myopically disregards the resource dynamics and the other is farsighted (does not disregard the resource dynamics), the farsighted party has a strong advantage over the myopic one. We also find the same outcome when the farsighted party leads operations by openly stating the course of its actions and the myopic party is a follower that responds to the leader party's actions.

## 12.2 The Problem

The available resource stock of a terrorist organization at time $t$, $X(t)$, arises from the operations of the two counteracting parties – the terrorists and the government:

$$dX(t) = \varepsilon X(t) + u_1(t) - u_2(t) + \sigma X(t)dW(t), \qquad X(0) = x_0, \tag{12.1}$$

where $u_1(t) \geq 0$ and $u_2(t) \geq 0$ represent the attack intensity of the terrorist and governmental counter-terror efforts, respectively; $W(t)$ is the standard Wiener process representing uncertainty associated with the resource stock, $\sigma$ represents the volatility of the resource stock, and $\varepsilon$ represents the resource natural growth rate (accumulation) in absence of actions.

Economic losses caused by terror attacks are measured as the quadratic cost of the terrorist resource stock $X^2(t)$ at each time $t$ along with the stock salvage value $aX^2(T)$ and the difference between the quadratic costs of the government and the terrorist actions $u_2^2(t) - u_1^2(t)$ over planned time horizon $T$:

$$J = E\left[\int_0^T \left(X^2(t) + u_2^2(t) - u_1^2(t)\right) dt + aX^2(T)\right]. \tag{12.2}$$

The terrorists aim to maximize expected damage (12.2), and the government aims to minimize it.

We assume that the terrorists' resource is not observable. The government does not have accurate information about the resources, and even the terrorist organization can only guess about its total resource stock. Therefore, we next assume that $X(t)$ is known only at $t = 0$ and cannot be observed until at least $t = T$ while the parties still plan to act. That is, the government and the terrorists apply deterministic commitment strategies $\{u_2(t) \geq 0 \lfloor X(0) = x_0, 0 \leq t \leq T\}$ and $\{u_1(t) \geq 0 \lfloor X(0) = x_0, 0 \leq t \leq T\}$ to deter/inflict the damage across all possible stochastic scenarios. Consequently, by denoting $E[X(t)] = x(t)$, we obtain the following from (12.1):

$$\frac{dx(t)}{dt} = \varepsilon x(t) + u_1(t) - u_2(t). \tag{12.3}$$

Using the Ito lemma, $d[X^2(t)] = 2[X(t)](\varepsilon X(t) + u_1(t) - u_2(t) + \sigma X(t)dW(t))dt + \sigma^2 X^2(t)$, $dE[X^2(t)] = 2E[[X(t)](\varepsilon X(t) + u_1(t) - u_2(t))]dt + \sigma^2 E[X^2(t)]dt$. Next, we introduce a new state variable, $y(t) = E[X^2(t)]$, and obtain its dynamics:

$$\frac{dy}{dt} = 2\left(u_1(t) - u_2(t)\right)x(t) + \left(2\varepsilon + \sigma^2\right)y(t), \tag{12.4}$$

where $y(0) = E\left[X^2(0)\right] = x_0^2$. Consequently, the stochastic differential game (12.1), (12.2) is reduced to the objective

$$J = \int_0^T \left(y(t) + u_2^2(t) - u_1^2(t)\right)dt + ay(T)\Big], \tag{12.5}$$

which the terrorist organization maximizes with the admissible control $\{u_1(t) \geq 0 \lfloor X(0) = x_0, 0 \leq t \leq T\}$ and the government minimizes with the admissible control $\{u_2(t) \geq 0 \lfloor X(0) = x_0, 0 \leq t \leq T\}$ subject to the dynamics of (12.3) and (12.4).

## 12.3 Equilibrium Conditions and Properties

The terrorist Hamiltonian $H_1$ and the government Hamiltonian $H_2$ are given by

$$\begin{aligned} H_1 =& y(t) + u_2^2(t) - u_1^2(t) + \lambda_1(t)\left(\varepsilon x(t) + u_1(t) - u_2(t)\right) \\ &+ \psi_1(t)\left(2\left(u_1(t) - u_2(t)\right)x(t) + \left(2\varepsilon + \sigma^2\right)y(t)\right), \end{aligned} \tag{12.6}$$

$$\begin{aligned} H_2 =& - y(t) - u_2^2(t) + u_1^2(t) + \lambda_2(t), \left(\varepsilon x(t) + u_1(t) - u_2(t)\right) \\ &+ \psi_2(t)\left(2\left(u_1(t) - u_2(t)\right)x(t) + \left(2\varepsilon + \sigma^2\right)y(t)\right), \end{aligned} \tag{12.7}$$

in which terrorist- and government-related co-states $(\psi_1(t)\lambda_1(t))$ and $(\psi_2(t)\lambda_2(t))$, respectively, are determined by co-state differential equations:

$$\begin{aligned}\dot{\lambda}_1(t) &= -\varepsilon\lambda_1(t) - 2\psi_1(t)\,(u_1(t) - u_2(t))\,;\,\dot{\lambda}_2(t)\\ &= -\varepsilon\lambda_2(t) - 2\psi_2(t)\,(u_1(t) - u_2(t))\,;\,\lambda_1(T) = \lambda_2(T) = 0;\end{aligned} \tag{12.8}$$

$$\dot{\psi}_1(t) = -1 - \left(2\varepsilon + \sigma^2\right)\psi_1(t);\ \dot{\psi}_2(t) = 1 - \left(2\varepsilon + \sigma^2\right)\psi_2(t);\ \psi_1(T) = -\psi_2(T) = a. \tag{12.9}$$

From (12.9), we observe that.
$\dot{\psi}_1(t) + \dot{\psi}_2(t) = -\left(2\varepsilon + \sigma^2\right)\left(\psi_1(t) + \psi_2(t)\right)$ and $\psi_1(T) + \psi_2(T) = 0$.
That is,

$$\psi_1(t) + \psi_2(t) = 0 \ \text{ and } \ \dot{\psi}_1(t) + \dot{\psi}_2(t) = 0\,. \tag{12.10}$$

We next omit the independent variable $t$ whenever dependence on time is obvious. Note that the explicit solution for (12.9) is

$$\begin{aligned}\psi_1 &= \frac{1}{2\varepsilon + \sigma^2}\left(e^{(2\varepsilon+\sigma^2)(T-t)}\left(1 + a\left(2\varepsilon + \sigma^2\right)\right) - 1\right) \quad \text{and}\\ \psi_2 &= \frac{1}{2\varepsilon + \sigma^2}\left(1 - e^{(2\varepsilon+\sigma^2)(T-t)}\left(1 + a\left(2\varepsilon + \sigma^2\right)\right)\right).\end{aligned} \tag{12.11}$$

From (12.10) and (12.11), we readily conclude:

**Lemma 12.1** $\psi_1 > 0$, $\psi_2 < 0$ and $\dot{\psi}_1 < 0$, $\dot{\psi}_2 > 0$ always hold. ■

An optimal response by each party is given by the first-order optimality conditions, which result in

$$u_1 = \begin{cases} \frac{\lambda_1 + 2\psi_1 x}{2}, \text{if } \lambda_1 \geq -2\psi_1 x \\ 0, if \text{ otherwise} \end{cases}, \quad \text{and} \quad u_2 = \begin{cases} \frac{-\lambda_2 - 2\psi_2 x}{2}, \text{if } \lambda_2 \leq -2\psi_2 x \\ 0, \text{if otherwise} \end{cases} \tag{12.12}$$

Let an equilibrium solution exist over the time horizon such that

$$-2\psi_1 \leq \frac{\lambda_1}{2x} \quad \text{and} \quad \frac{\lambda_2}{2x} \leq -2\psi_2. \tag{12.13}$$

We next show the following property:

**Lemma 12.2** Let (12.12) hold. Consider a symmetric equilibrium, then $\lambda_1 = \lambda_2 = 0$.

**Proof** We start from an interior solution, $2(u_1 - u_2) = \lambda_1 + 2\psi_1 x + \lambda_2 + 2\psi_2 x$ (see 12.12). Substituting the interior solution from (12.12) into (12.8) leads to

$$\dot{\lambda}_1 = -\varepsilon\lambda_1 - \psi_1(\lambda_1 + 2\psi_1 x + \lambda_2 + 2\psi_2 x);\ \dot{\lambda}_2 = -\varepsilon\lambda_2 - \psi_2(\lambda_1 + 2\psi_1 x + \lambda_2 + 2\psi_2 x), \quad (12.14)$$

which, with respect to (12.10), results in

$$\dot{\lambda}_1 = -\varepsilon\lambda_1 - \psi_1(\lambda_1 + \lambda_2);\ \dot{\lambda}_2 = -\varepsilon\lambda_2 - \psi_2(\lambda_1 + \lambda_2). \quad (12.15)$$

Recalling that $\psi_1 + \psi_2 = 0$ (see (12.10)), we obtain:

$$\dot{\lambda}_1 + \dot{\lambda}_2 = -\varepsilon(\lambda_1 + \lambda_2). \quad (12.16)$$

By accounting for the transversality conditions from (12.8), we find that the only solution for (12.16) is $\lambda_1 + \lambda_2 = 0$ and, therefore, that

$$\dot{\lambda}_1 + \dot{\lambda}_2 = 0. \quad (12.17)$$

Consequently, $\lambda_1 + 2\psi_1 x + \lambda_2 + 2\psi_2 x = 0$ and, from (12.14), $\dot{\lambda}_1 = -\varepsilon\lambda_1$ and $\dot{\lambda}_2 = -\varepsilon\lambda_2$, which, with respect to the transversality conditions from (12.8), results in $\lambda_1 = \lambda_2 = 0$.

Now consider a boundary solution $u_1 = u_2 = 0$. In that case, (12.8) again transforms into $\dot{\lambda}_1 = -\varepsilon\lambda_1$; $\dot{\lambda}_2 = -\varepsilon\lambda_2$, and, therefore, (12.17) holds and $\lambda_1 = \lambda_2 = 0$. ■

Note that the found co-state property ensures that the optimality conditions in (12.12) are always met for an interior solution and are not met for boundary equilibrium solution $u_1 = u_2 = 0$.

We next verify that a non-symmetric equilibrium does not exist.

**Lemma 12.3** Let $u_1 > 0$ and $u_2 = 0$ or let $u_2 > 0$ and $u_1 = 0$. Then, the optimality conditions in (12.12) do not hold.

**Proof** The proof is by contradiction. Assume, for example, that the equilibrium solution is $u_1 > 0$ and $u_2 = 0$. Then, $\frac{\lambda_2}{2x} \geq -2\psi_2$ must hold. Co-state Eq. (12.8) takes the following form:

$$\dot{\lambda}_1 = -\varepsilon\lambda_1 - \psi_1(\lambda_1 + 2\psi_1 x),\ \lambda_1(T) = 0 \quad \text{and}$$
$$\dot{\lambda}_2 = -\varepsilon\lambda_2 - \psi_2(\lambda_1 + 2\psi_1 x),\ \lambda_2(T) = 0 \quad (12.18)$$

and we observe that (12.17) once again holds. From $u_1 > 0$, $\lambda_1 + 2\psi_1 x > 0$, and from (12.18) and the transversality conditions, $\lambda_1 > 0$. Since $\lambda_1 = -\lambda_2$, we find that $\lambda_2 < 0$. And since $\psi_2 < 0$, we readily observe that $\frac{\lambda_2}{2x} \geq -2\psi_2$ cannot hold, which contradicts our assumption that $u_1 > 0$ and $u_2 = 0$ meets the optimality conditions (12.12). Similarly, it is readily verified that $u_1 = 0$ and $u_2 > 0$ does not meet (12.12). ■

From Lemmas 1 through 3, we conclude:

**Theorem 12.1** The differential game specified in (12.3) through (12.5) between a farsighted government and a terrorist organization has only an interior open-loop Nesh equilibrium that is always unique and symmetric and is given by.

$$u_1 = u_2 = \frac{1}{2\varepsilon + \sigma^2} \left( e^{(2\varepsilon+\sigma^2)(T-t)} \left( 1 + a \left( 2\varepsilon + \sigma^2 \right) \right) - 1 \right) x. \blacksquare \tag{12.19}$$

Note that Eq. (12.19) is a feedback representation of the open-loop Nash equilibrium solution, which is based on the expected terrorist resource stock rather than on the true resource stock. Furthermore, the greater the expected resource stock *x*, the greater the actions undertaken by the parties, leading to exponential growth in the resource stock. Specifically, since the equilibrium is symmetric,

$$x = x(0)e^{\varepsilon t}, \tag{12.20}$$

which transforms (12.19) into the explicit open-loop equilibrium solution:

$$u_1 = u_2 = \frac{1}{2\varepsilon + \sigma^2} \left( e^{(2\varepsilon+\sigma^2)(T-t)} \left( 1 + a \left( 2\varepsilon + \sigma^2 \right) \right) - 1 \right) x(0)e^{\varepsilon t}. \tag{12.21}$$

From (12.21), we observe that

$$\begin{aligned} u_1(0) =& u_2(0) = \frac{1}{2\varepsilon + \sigma^2} \left( e^{(2\varepsilon+\sigma^2)T} \left( 1 + a \left( 2\varepsilon + \sigma^2 \right) \right) - 1 \right) x(0) \text{ and} \\ u_1(T) =& u_2(T) = ax(0)e^{\varepsilon T}. \end{aligned} \tag{12.22}$$

That is, given fixed planning horizon *T*, the greater the value of the initial resource stock *ax*(0) and/or the associated rate of growth $\varepsilon$ per stock and time unit, the stronger the actions undertaken by the parties by the end of the planned period. The effect of uncertainty $\sigma$ is of special interest and is discussed in the numerical analysis.

## 12.4 Myopic Policies

### *12.4.1 Ignoring Dynamics in Variation of Resources by both Parties*

Since $y = E[X^2] = Var[X] + E[X]^2$, ignoring the dynamics of *y* is associated with ignoring the variance of the resource stock. This myopic behavior implies that $\psi_1 = \psi_2 = 0$. Then, from (12.14), we have $\dot{\lambda}_1 = -\varepsilon\lambda_1, \dot{\lambda}_2 = -\varepsilon\lambda_2$; that is, $\lambda_1 = \lambda_2 = 0$, $u_1 = 0$, $u_2 = 0$ (see (12.13)), and $\dot{x} > 0$. Neither the terrorist organization nor the government does anything to change the current situation, which indicates that the terrorist's resources will increase.

**Proposition 12.1** If both parties ignore variation in the resource dynamics (behave myopically), there will be no terror acts and no government anti-terrorist activities, and expected terrorist resources will increase the same amount as under farsighted conditions (according to (12.20) when dynamics are not ignored). ∎.

Since the parties take no actions while the resource stock evolves the same as under farsighted conditions, this myopic behavior enables both parties to improve their objective functions (payoffs).

### 12.4.2 One Party Ignores Variation in Resource Dynamics

Equation (12.11) describes the case in which one party is myopic while the other is farsighted. When the government is myopic in terms of variation in resource stock dynamics while the terrorists are farsighted, $\psi_1$ is given by (12.11) and $\psi_2 = 0$. Then, according to (12.8), $\dot{\lambda}_2 = -\varepsilon\lambda_2$, which again results in $\lambda_2 = 0$ and $u_2 = 0$. Consequently, $\dot{\lambda}_1 = -\varepsilon\lambda_1 - 2\psi_1 u_1$. When taking (12.12) and (12.3) into account, $\dot{x} = \varepsilon x + u_1$ leads to a system of two differential equations with two unknowns:

$$\dot{\lambda}_1 = -\varepsilon\lambda_1 - \psi_1\left(\lambda_1 + 2\psi_1 x\right) \quad \text{where } \psi_1 \text{ is given by (12.9) and } \dot{x} = \varepsilon x + \frac{\lambda_1 + 2\psi_1 x}{2}. \tag{12.23}$$

The solution of this two-point boundary-value problem, ($X(0) = x_0$ and $\lambda_1(T) = 0$), determines $x$ and $\lambda_1$; therefore, equilibrium control $u_1 = \frac{\lambda_1 + 2\psi_1 x}{2}$. Though we solve this system only numerically, insights can be obtained analytically by comparing two games, one characterized by farsighted parties (upper index $f$ distinguishes between the games) and the other by one farsighted party (index *fm*) and one myopic party (index *m*). Specifically, we consider the difference in behaviors of the farsighted parties in two games:

$$2u_1^{fm} - 2u_1^{f} = \lambda_1^{fm} + 2\psi_1\left(x^{fm} - x^{f}\right) \tag{12.24}$$

where $\psi_1 = \psi_1^{fm} = \psi_1^{f}$ and

$$\dot{x}^{fm} = \varepsilon x^{fm} + u_1^{fm}, u_2^{m} = 0 \quad \text{and} \quad \dot{x}^{f} = \varepsilon x^{f}, u_1^{f} = u_2^{f}. \tag{12.25}$$

Given that $X(0) = x_0$ in both games, we straightforwardly conclude from (12.24) and (12.25) that $x^{fm} = x^{f} + \int_0^T u_1^{fm} dt$. Since $\lambda_1^{m}(T) = 0$ when $\psi_1 = a$ and given that $\lambda_1^{fm}(0) > 0$, then, from (12.18), we readily find that $\dot{\lambda}_1^{fm} < 0$ over the entire time horizon and, thus, that $\lambda_1^m > 0$, $u_1^m > 0$. Consequently, a sufficiently small $\lambda_1^{fm}(0) > 0$ always exists, ensuring that $\lambda_1^{fm}(T) = 0$ is met regardless of how short time horizon is. That is, there is always a feasible solution when $\lambda_1^{fm} > 0$ and, therefore, $2u_1^{fm} - 2u_1^f = \lambda_1^{fm} + 2\psi_1 \left(x^{fm} - x^f\right) = \lambda_1^{fm} + 2\psi_1 \int_0^T u_1^{fm} dt > 0$.

If the terrorist organization is myopic and the government is farsighted, then, using similar arguments, we find that $\psi_1 = 0$ and $\psi_2$ is given by (12.11), which results in $\lambda_1 = 0$ and $u_1 = 0$. Next, comparing these results to results when both parties are farsighted, we find that $2u_2^{mf} - 2u_2^f = -\lambda_2^{mf} - 2\psi_2 \left(x^{mf} - x^f\right)$, $\dot{x}^{mf} = \varepsilon x^{mf} - u_2^{mf}$, and $x^{mf} = x^f - \int_0^T u_2^{mf} dt$. Therefore, from $\psi_2 < 0$ and $\lambda_2^m > 0$, we obtain $2u_2^{mf} - 2u_2^f = -\lambda_2^{mf} + 2\psi_2 \int_0^T u_2^{mf} dt < 0$ using the same arguments. Summarizing our findings:

**Proposition 12.2** When one party is myopic and the other is farsighted, the myopic party does not take action (anti-terror or terror); the farsighted party acts, engaging in more terror/less anti-terrorism than it conducts when both parties are either farsighted or myopic. Furthermore, myopic terrorist resources can decrease in response to a farsighted government. If the resource salvage value $a$ is large relative to the resource accumulation rate, $\varepsilon$, farsighted terrorist resources increase more quickly under the myopic/farsighted split than when both parties are either farsighted or myopic. ■

Figure 12.1 illustrates Proposition 2 for the three games: (a) *f* when both parties farsighted, (b) *fm* when the terrorists are farsighted and the government is myopic, and (c) *mf* when the terrorists are myopic and the government is farsighted.

From Fig. 12.1, we observe that, when both parties are farsighted, $u_1^{fm}(t) > u_1^f(t) > u_2^{mf}(t)$ for $t \in [0, T]$ (panel 1a), activities of both parties decrease monotonically over time. A myopic government leads to the greatest activity and

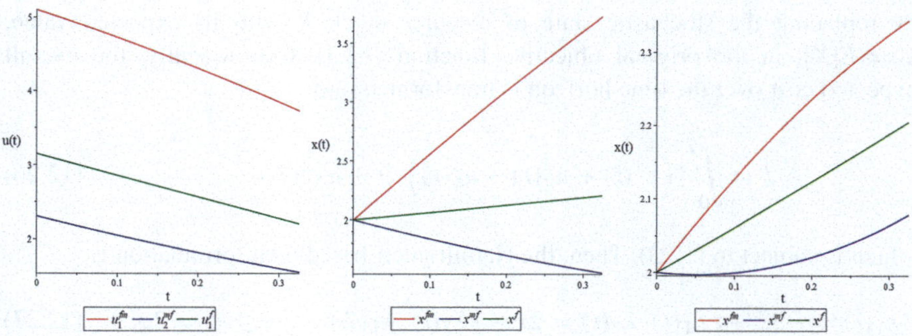


(a) (b) (c)

**Fig. 12.1** Control and state dynamics when $\varepsilon = 0.3$, $\sigma = 0.05$, $x(0) = x_0 = 2$, and $T = 0.325$. (**a**) $u(t)$ both farsighted, (**b**) $x(t)$ for $a = 1$ farsighted terrorist, (**c**) $x(t)$ for $a = 0.01$ farsighted government

the greatest accumulation of resources by farsighted terrorists (panel 1b). Myopic terrorists lead to less action by the farsighted government so the myopic terrorist resource decreases when $a$ is high ($a = 1$, panel 1b) and decreases only initially when $a$ is low ($a = 0.01$, panel 1c).

We next assume leader-follower relationships between the parties.

### 12.4.3 Farsighted Leader and Myopic Follower Ignore Dynamics in Resource Variation

Assume that one party is myopic (e.g., the government) in terms of variation in resource dynamics. That implies that $\psi_2 = 0$ when the other party is farsighted and leading (openly stating its planned actions). The follower's solution is $u_2 = 0$ because we once again observe that $\lambda_2 = 0$, and the Stackelberg leader's solution once again leads to $\dot{\lambda}_1 = -\varepsilon\lambda_1 - \psi_1(\lambda_1 + 2\psi_1 x)$ in which $\psi_1$ is determined by (12.11). We therefore conclude:

**Proposition 12.3** Assume one party is farsighted (e.g., the terrorist organization) and is leading the interaction by openly stating the course of its actions for the entire time horizon at $t = 0$ while the other party is myopic (e.g., the government) and responds to the leader's actions. Then, the Stackelberg solution is identical to the Nash solution defined by (12.23) when the same parties are farsighted and myopic.
■

### 12.4.4 Myopic Parties that Ignore Stochastics of the Resources

In this case, we assume that the existence of variance is simply ignored by the two myopic parties as a factor affecting their objective functions. This is accomplished by replacing the stochastic state of resource stock $X$ with its expected value, $x = E[X]$, in the original objective function (12.3). Consequently, the overall expected cost over the time horizon is transformed into

$$J = \int_0^T \left(x^2(t) + u_2^2(t) - u_1^2(t)\right) dt + ax^2(T), \tag{12.26}$$

which is subject to (12.3). Then, the Hamiltonian-based dual formulation is.

$$\dot{\lambda}_1(t) = -2x - \varepsilon\lambda_1(t), \dot{\lambda}_2(t) = 2x - \varepsilon\lambda_2(t), \lambda_1(T) = -\lambda_2(T) = 2a; \tag{12.27}$$

$$u_1 = \begin{cases} \frac{\lambda_1}{2}, if\ \lambda_1 \geq 0 \\ 0, if \text{ otherwise} \end{cases}, u_2 = \begin{cases} \frac{-\lambda_2}{2}, if\ \lambda_2 \leq 0 \\ 0, if \text{ otherwise} \end{cases}. \tag{12.28}$$

It then is straightforward to see that we again have $u_1 = u_2$, $x = x(0)e^{\varepsilon t}$, and $\dot{\lambda}_1(t) = -2x(0)e^{\varepsilon t} - \varepsilon\lambda_1(t)$; that is, $\lambda_1 = \frac{x(0)}{\varepsilon}e^{\varepsilon t}\left(e^{2\varepsilon(T-t)} - 1\right) + 2ae^{\varepsilon(T-t)} = -\lambda_2$. Consequently, the equilibrium control is given by

$$u_1 = u_2 = \frac{x(0)}{2\varepsilon}e^{\varepsilon t}\left(e^{2\varepsilon(T-t)} - 1\right) + ae^{\varepsilon(T-t)}. \tag{12.29}$$

Note that, by Theorem 1, the commitment Nash equilibrium is characterized by the shadow price of the expected resource state dynamic being zero while equilibrium control (12.29) is characterized by the shadow price of the expected resource state dynamic not being ignored (non-zero). And unlike in the original stochastic game, ignoring the state dynamics of one of the parties (defined by setting the co-state at zero in (12.27)) does not affect the behavior of the other party. We again find, as in the previous results, that the farsighted party always has an advantage over the myopic party.

Recalling (12.21), we conclude that:

**Proposition 12.4** When $\frac{1}{2\varepsilon+\sigma^2}\left(e^{(2\varepsilon+\sigma^2)(T-t)}\left(1 + a\left(2\varepsilon + \sigma^2\right)\right) - 1\right)x(0)e^{\varepsilon t} > \frac{x(0)}{2\varepsilon}e^{\varepsilon t}\left(e^{2\varepsilon(T-t)} - 1\right) + ae^{\varepsilon(T-t)}$, the equilibrium actions determined by Theorem 1 for both parties exceed the actions they take when their objective functions are as in (12.26). Otherwise, the parties take less action. ■

Proposition 4 is illustrated in Fig. 12.2 for the original data (panel a) and for a new set of data (panel b). Panel b of the figure shows that intersection between actions of farsighted parties, $u_1^f$, and actions by parties ignoring variability (stochastic nature),

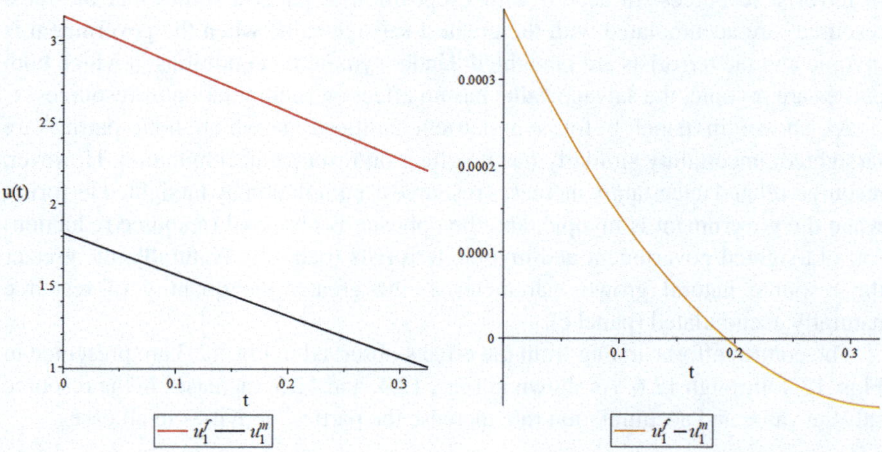


(a) (b)

**Fig. 12.2** Dynamics of (**a**) $u_1^{fm}$ and $u_1^m$ farsighted terrorist and ignoring variation under the original data set and (**b**) Difference between $u_1^{fm}$ and $u_1^m$ from new data: $\sigma = 0.09$, $\varepsilon = 0.3$, $a = 0.01$, $x_0 = 0.9$, $T = 0.325$

$u_1^m$, is possible, but the differences between the controls are tiny, indicating that the actions remain nearly the same over the planning horizon. From the original data used to illustrate Proposition 1 (panel a), we find that farsighted parties are always more active than myopic parties.

## 12.5 Numerical Analysis

The numerical analysis compares the actions of the terrorists and the government when one party is myopic and the other is farsighted for various data. In addition, we determine the effect of the problem parameters on each party's objective function, $J$ (maximization of terrorist economic damage/anti-terrorism efforts that reduce terrorist economic damage (12.5)). The basic (initial) data used for the analysis is presented in Table 12.1. Recall that we use $u_1^{fm}$ and $u_2^{mf}$ to indicate actions taken when the terrorists and the government, respectively, are farsighted and the other party is myopic and that $u_1^{fm}$ and $u_2^{mf}$ are determined by (12.12) and (12.23), respectively. In turn, $u_1^f$ reflects the actions taken when both sides are farsighted and is calculated according to (12.21).

Note that parameter $a$ in Table 12.1 reflects economic damage potentially suffered by the government from each unit remaining in the terrorists' stock at the end of the planning horizon while $\varepsilon$ indicates the natural accumulation of the terrorists' resource stock. Uncertainty associated with the terrorist resource is measured by the resource's volatility, $\sigma$. The parameters in Table 12.1, $x_0$ and $T$, denote the initial resource stock and the planning horizon, respectively.

Panels a, b, and c in Fig. 12.3 show the effect of $a$, $\sigma$, and $\varepsilon$ on the terminal stock of terrorist resources. In accord with Proposition 2, panel a shows that the most resources are accumulated with the greatest salvage value when the government is myopic and the terrorists are farsighted. Under symmetric conditions in which both parties are myopic, the salvage value has no effect on accumulation of resources.

As shown in panel b for symmetric conditions in which both parties are farsighted, uncertainty similarly has no effect on resource accumulation. However, resource-related uncertainty increases resource accumulation by farsighted terrorists when the government is myopic, and the opposite is observed (resource reduction) for a farsighted government and myopic terrorists (panel b). Naturally, the greater the resource natural growth parameter $\varepsilon$, the greater the quantity of resource naturally accumulated (panel c).

The control efforts arising from the effects observed in Fig. 12.3 are presented in Figs. 12.4. through 12.6. As shown in Figs. 12.4. and 12.5, increases in the resource salvage value and accumulation rate increase the parties' activities in all cases.

**Table 12.1** Dataset

| $a$ | $\varepsilon$ | $\sigma$ | $x_0$ | $T$ |
|---|---|---|---|---|
| 1 | 0.3 | 0.05 | 2 | 0.325 |

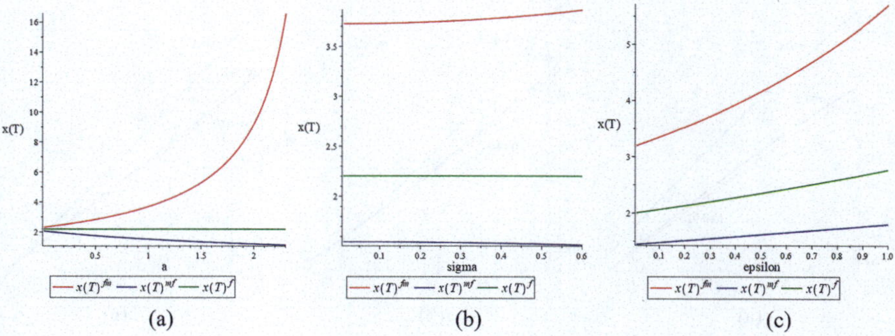


**Fig. 12.3** Effects of resource salvage value, resource uncertainty, and natural resource accumulation on the terminal level of the resource $x(T)$. (**a**) $a$ economic damage, (**b**) $\sigma$ resource uncertainty, (**c**) $\varepsilon$ natural resource accumulation

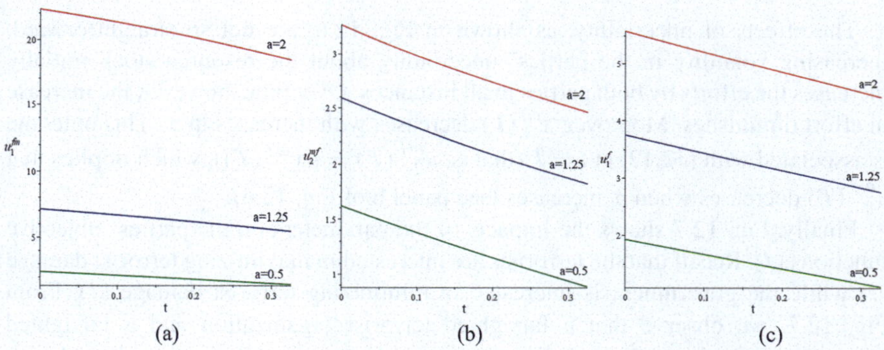


**Fig. 12.4** Control efforts arising from variation in resource salvage value $a$. (**a**) $u_1^{fm}$ farsighted terrorist, (**b**) $u_2^{mf}$ farsighted government, (**c**) $u_1^{f}$ both farsighted

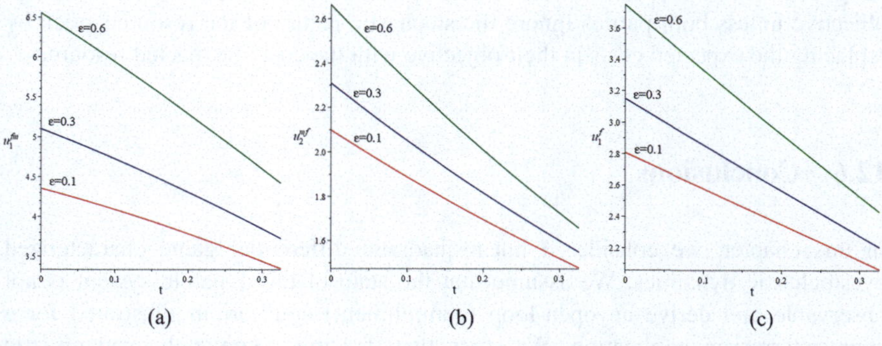


**Fig. 12.5** Control efforts arising from variation in natural resource accumulation $\varepsilon$. (**a**) $u_1^{fm}$ farsighted terrorist, (**b**) $u_2^{mf}$ farsighted government, (**c**) $u_1^{f}$ both farsighted

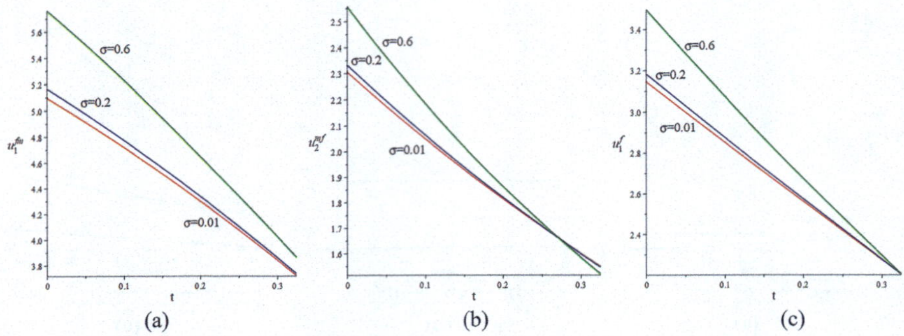


**Fig. 12.6** Control efforts arising from variation in resource stock uncertainty $\sigma$. (**a**) $u_1^{fm}$ farsighted terrorist, (**b**) $u_2^{mf}$ farsighted government, (**c**) $u_1^f$ both farsighted

The effects of uncertainty, as shown in Fig. 12.6, are not so straightforward. Increasing volatility in the parties' uncertainty about the resource stock initially increases the efforts by both parties in all instances. Over time, however, the increase in effort diminishes. Moreover, $x^{mf}(T)$ decreases with increases in $\sigma$. This outcome is associated with (12.12) at $t = T$ (that is, $u_2^{mf}(T) = ax^{mf}(T)$), which implies that $u_2^{mf}(T)$ decreases when $\sigma$ increases (see panel b of Fig. 12.6).

Finally, Fig. 12.7 shows the impacts of the parameters on the parties' objective functions ($J$). Recall that the terrorists are interested in maximizing terrorist damage ($J$) while the government is interested in minimizing terrorist damage ($J$). From Fig. 12.7, we observe that a farsighted terrorist organization and a farsighted government are always better off regardless of the farsighted/myopic policy chosen by the other party. Moreover, as the values of the problem parameters increase, the economic damage inflicted by the terrorist organization increases, thus decreasing the government's objective. Thus, we find that uncertainty improves the terrorists' objective unless both parties ignore the stochastic nature of the resource stock by replacing the expected costs in their objective with the cost of expected resources.

## 12.6 Conclusions

In this chapter, we consider a linear-quadratic differential game characterized by stochastic dynamics. We assume that the state of the dynamic system is not observable and derive an open-loop (commitment) equilibrium illustrated for a counter-terrorism application. We show that, for this game, only a symmetric open-loop equilibrium exists, and its feedback representation is based on expected terrorist resources rather than on the true unobservable state of those resources. The greater the expected stock of the resource, the greater the actions the parties undertake, and the resource stock grows exponentially.

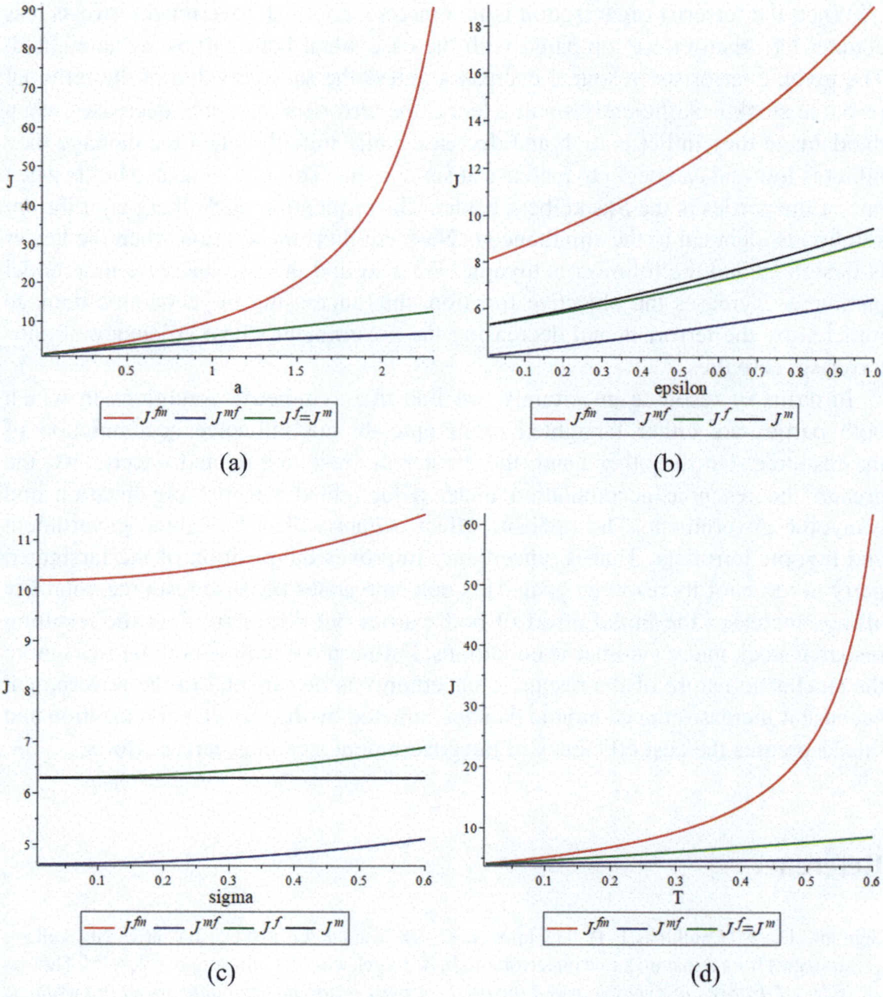


**Fig. 12.7** Influence of variation in the parameters on the parties' objective functions $J$. **(a)** $a$ resource salvage value, **(b)** $\varepsilon$ natural resource accumulation, **(c)** $\sigma$ resource uncertainty, **(d)** $T$ time horizon

We find that both parties are better off in terms of their objective functions when they are myopic and ignore variations in the resource. This occurs because the resource stock in that case evolves in exactly the same way as when both parties are farsighted. No actions are exerted so no control-related costs are incurred.

We further find that a farsighted party always has the advantage over a myopic party. For example, with a myopic government and a farsighted terrorist organization, the terrorists act but the government does not. Consequently, the greatest terrorist activity and accumulation of resources occurs in this case.

When the terrorist organization is myopic, a farsighted government invests less counter-terrorism effort compared with the case when both parties are farsighted. The myopic terrorists' resource decreases unless the salvage value of the terminal resource stock is sufficiently small. Hence, the terrorists' resource decreases when the damage they inflict is high and decreases only initially when the damage they inflict is low and can begin to increase at some point. This outcome also holds when one of the parties is the Stackelberg leader. The sequential Stackelberg equilibrium solution is identical to the simultaneous Nash equilibrium solution when the leader is farsighted and the follower is myopic. We also find that any increase in a model parameter increases the objective function, thus increasing the economic damage inflicted by the terrorists and decreasing the government's cost efficiency of anti-terrorism activities.

In terms of resource uncertainty, we find that symmetric conditions in which both parties are either farsighted or myopic do not influence accumulation of the resource. On the other hand, the greater the resource-related uncertainty, the greater the resource accumulation under a farsighted terrorist organization and a myopic government. The opposite effect occurs with a farsighted government and myopic terrorists. That is, uncertainty improves the position of the farsighted party in terms of its resource goal. This outcome arises because resource volatility always increases the initial effort of both parties but does not affect the resulting resource stock under symmetric conditions. Furthermore, unless both parties ignore the stochastic nature of the resource, uncertainty is detrimental to the government because it increases the economic damage inflicted by the terrorist organization and thus decreases the cost efficiency of the government's counter-terror efforts.

## References


Behrens, D. A., Caulkins, J. P., Feichtinger, G., & Tragler, G. (2007). Incentive Stackelberg strategies for a dynamic game on terrorism. In S. Jørgensen, M. Quincampoix, & V. L. Thomas (Eds.), *Advances in dynamic game theory. In Annals of the international society of dynamic games* (Vol. 9, pp. 459–486). Birkhäuser.

Bismut, J. M. (1976). Linear quadratic optimal stochastic control with random coefficients. *SIAM Journal on Control and Optimization, 14*, 419–444.

Chen, S., & Yong, J. (2001). Stochastic linear quadratic optimal control problems. *Applied Mathematics and Optimization, 43*, 21–45. https://doi.org/10.1007/s002450010016

Crettez, B., & Hayek, N. (2014). Terrorist eradication versus perpetual terror war. *Journal of Optimization Theory and Applications, 160*(2), 679–702.

Feichtinger, G., & Novak, A. (2008). Terror and counter terror operations: Differential game with cyclical Nash solution. *Journal of Optimization Theory and Applications, 139*(3), 541–556.

Kaplan, E. H. (2010). Terror queues. *Operations Research, 58*(4-part-1), 773–784.

Kaplan, E. H. (2015). Socially efficient detection of terror plots. *Oxford Economic Papers, 67*(1), 104–115.

Kogan, K., & Chernonog, T. (2019). Competition under industry-stock-driven prevailing market price: Environmental consequences and the effect of uncertainty. *European Journal of Operational Research, 276*(3), 929–946.

Li, B., Zhang, R., Jin, T., & Shu, Y. (2021). Parametric approximate optimal control of uncertain differential game with application to counter terror. *Chaos, Solitons and Fractals, 146*, 110940.

Li, N., Li, X., Peng, J., & Xu, Z. Q. (2022). Stochastic linear quadratic optimal control problem: A reinforcement learning method. *IEEE Transactions on Automatic Control, 67*(9), 5009–5016.

Seidl, A., Kaplan, E. H., Caulkins, J. P., Wrzaczek, S., & Feichtinger, G. (2016). Optimal control of a terror queue. *European Journal of Operational Research, 24*, 246–256.

Sun, J., & Yong, J. (2020). *Stochastic linear-quadratic optimal control theory: Differential games and mean-field problems*. Springer.

Sun, Y., Ding, C., & Zhu, Y. (2018). Saddle point equilibrium under uncertain environment. *Math Method Appl Sci, 41*(5), 2063–2073.

Tang, S. (2020). *Stochastic linear-quadratic control in encyclopedia of systems and control*. In Baillieul, J., & Samad, T. (Eds.). Springer, pp. 1–6.

Wei, Q., J. Yong, & Z. Yu (2019). *Linear quadratic stochastic optimal control problems with operator coefficients: Open-loop solutions, ESAIM: COCV*, p. 25. doi:https://doi.org/10.1051/cocv/2018013.

Yong, J., & Zhou, X. Y. (1999). *Stochastic controls: Hamiltonian systems and HJB equations*. Springer.

Zhuang, J., Bier, V. M., & Alagoz, O. (2010). Modeling secrecy and deception in a multiple-period attacker–defender signaling game. *European Journal of Operational Research, 203*(2), 409–418.

The manufacturer's authorised representative in the EU is Springer Nature Customer Service Centre GmbH, Europaplatz 3, 69115 Heidelberg, Germany. If you have any concerns regarding our products, please contact ProductSafety@springernature.com

Printed and bound by CPI Group (UK) Ltd, Croydon, CR0 4YY
05/10/2026
02228130-0004